THE 388TH TACTICAL FIGHTER WING

LIBERTAS VELMORS

AT KORAT ROYAL THAI AIR FORCE BASE 1972

Don Logan

Schiffer Military/Aviation History
Atglen, PA

ACKNOWLEDGEMENTS

I would like to thank the following individuals who have helped me in this project: Tom Brewer, Joe Bruch, Michael France, Jerry Geer, Vic Hilgren, Tom Kaminski, Craig Kaston, Patrick Martin, Charles T. Robbins, Brian C. Rogers, Mick Roth, and Douglas Slowiak. The unit patches used in this book were supplied by John Cook. The maps were drawn by Roger Johansen. Photographs, unless noted otherwise, were taken by the author. Some text in this book was previously contained in two articles written by the author for *The Journal of Military Aviation*.

THE AUTHOR

After graduating from California State University-Northridge with a BA degree in History, Don Logan joined the USAF in August of 1969. He flew as an F-4E Weapon Systems Officer (WSO), stationed at Korat RTAFB in Thailand, flying 133 combat missions over North Vietnam, South Vietnam, and Laos before being shot down over North Vietnam on July 5, 1972. He spent nine months as a POW in Hanoi, North Vietnam. As a result of missions flown in Southeast Asia, he received the Distinguished Flying Cross, the Air Medal with twelve oak leaf clusters, and the Purple Heart. After his return to the U.S., he was assigned to Nellis AFB where he flew as a rightseater in the F-111A. He left the Air Force at the end of February 1977.

In March of 1977 Don went to work for North American Aircraft Division of Rockwell International, in Los Angeles, as a Flight Manual writer on the B-1A program. He was later made editor of the Flight Manuals for B-1A #3 and B-1A #4. Following the cancellation of the B-1A production, he went to work for Northrop Aircraft as a fire control and ECM systems maintenance manual writer on the F-5 program.

In October of 1978 he started his employment at Boeing in Wichita, Kansas as a Flight Manual/Weapon Delivery manual writer on the B-52 OAS/CMI (Offensive Avionics System/Cruise Missile Integration) program. He is presently the editor for Boeing's B-52 Flight and Weapon Delivery manuals, B-1 OSO/DSO Flight Manuals and Weapon Delivery Manuals, and T-43A flight Manuals.

Don Logan is also the author of *Rockwell B-1B: SAC's Last Bomber* and *Northrop's T-38 Talon: A Pictorial History* (both available from Schiffer Publishing Ltd.).

Book Design by Robert Biondi

Copyright © 1995 by Don Logan.
Library of Congress Catalog Number: 95-67627

Printed in China.
ISBN: 0-88740-798-6

We are interested in hearing from authors with book ideas on related topics.

Published by Schiffer Publishing Ltd.
77 Lower Valley Road
Atglen, PA 19310
Please write for a free catalog.
This book may be purchased from the publisher.
Please include $2.95 postage.
Try your bookstore first.

CONTENTS

THE 388TH TACTICAL FIGHTER WING
KORAT ROYAL THAI AIR FORCE BASE 1972

INTRODUCTION

Following Desert Storm the U.S. Air Force embarked upon a dramatic re-organization. As part of this restructuring, the Air Force began forming composite wings. One of these, the 366th Wing at Mountain Home AFB, Idaho was to be the "air intervention wing." This wing, made up of various types of aircraft, was designed to assemble an integrated strike force at one base under a single wing.

This is not a new concept. Composite wings existed during World War II, the Korean conflict, and the war in Vietnam. One such composite wing taking part in the Vietnam War was the 388th Tactical Fighter Wing.

In 1972, during the Vietnam conflict, the 388th Tactical Fighter Wing was stationed at Korat Royal Thai Air Force Base in Thailand. The 388th TFW was made up of F-4E, F-105G, and EB-66 squadrons; in addition there were detachments of EC-121s, ABCCC C-130Es, HC-130Ps, F-4Ds, F-4C Wild Weasels, and A-7Ds. This aerial armada flew in support of LINEBACKER bombing operations in North Vietnam. In addition, as part of Young Tiger (the nickname under which SAC's KC-135As performed the mission of refueling tactical aircraft in Southeast Asia), KC-135As were assigned to Korat. This co-location of resources allowed for the planning and execution of specialized missions using the resources available.

For part of my USAF career I was stationed at Korat RTAFB. Flying from Korat, as an F-4E Weapon Systems Officer (WSO), I recorded 133 combat missions over North and South Vietnam and Laos from November of 1971 until July 5, 1972, when I was shot down over North Vietnam.

This book describes in both words and pictures the operations of the 388th Tactical Fighter Wing at Korat, during 1972, the time frame when I was assigned there. The majority of photographs in this book were taken by me in 1972 while I was stationed at Korat — these are left uncredited in the captions. Photographs by others are noted.

THE BASE

In April 1962, the U.S. Air Force had one officer and 14 airmen assigned to Korat RTAFB on temporary duty status. By 1972 there were are over 4,000 officers and airmen assigned at Korat, frequently described by visitors and assigned military personnel as "the best base in Southeast Asia."

The base is located about five miles south of Korat City, the third largest city in Thailand, with a population of 80,000. It is approximately 145 miles north and slightly east of Bangkok, Thailand's capital.

The air base grew from a little clearing in the jungle into a sprawling installation dotted with single and two-story barracks, modern maintenance facilities and well-designed offices. The paved streets and almost three miles of sidewalks made life comfortable for Korat's residents. Swimming pools, tennis courts, a golf driving range, a well equipped gymnasium and a new theater were only a part of a busy and varied recreation program.

Korat RTAFB was headquarters for the Royal Thai Air Force 3rd Helicopter Wing in addition to the 388th Tactical Fighter Wing, 553rd Reconnaissance Wing and over 30 associate units representing seven U.S. Air Force commands.

The field elevation of Korat is only 600 feet. Korat's single runway, 06/24 was 9845 feet long and 148 feet wide. The runway had a 1,000 foot overrun at both ends with barrier arresting cables crossing 2,000 feet from each end. A single taxiway ran the full length of the runway on the southeast side.

The aircraft parking ramp was also on the east side. EB-66s, C-130s, and EC-121s were parked on the northeast end of the ramp. The northeast end parking spaces were wide enough for one EC-121 or two EB-66s and had revetment walls between the parking spaces. The control tower, fire/rescue helicopter pad, and transient parking were at the center of the ramp.

Two rows of revetments for the F-4s were southwest of the control tower running parallel to the taxiway. Only the first row of F-4s had full, single aircraft, three side revetments. The revetments in the second F-4 parking row were two sided which allowed the aircraft to taxi in to and out of the revetments.

The F-105Gs parked on the open ramp between the F-4 revetments and the Air Terminal located on the southwest end of the ramp. There were no revetments on the southwest ramp for the F-105Gs, though they were sometimes parked in the F-4 revetments. The A-7s replaced the F-4s parked in the second row taxi through revetments. A parking pad of PSP (Perforated Steel Planking) was located on the runway side of the taxiway across from the F-105G parking.

Crewmembers at Korat lived in "hootches" like these. Each hootch had four two man bedrooms and a central bathroom.

This view looking southwest shows the base complex with the ramp and runway in the background.

THE ORGANIZATION

Thirteenth Air Force

The 388th Tactical Fighter Wing, when based at Korat, was part of the Thirteenth Air Force. The Thirteenth was based at Clark AB, Republic Of The Philippines during the Vietnam conflict and was part of the Pacific Air Forces (PACAF) based at Hickam AFB, Hawaii.

The command first was tagged with the unlucky number 13 back on January Thirteenth (naturally) in the year 1943. Its first home was on a tiny island in the South Pacific-New Caledonia. The world was engulfed by war, and the newly formed Thirteenth Air Force was to play a large part in the Pacific air war during World War II. The first commander, Major General Nathan F. Twining, once said the organization began "under difficult circumstances and in the face of an advancing enemy." (Despite his close association with the number "Thirteen," General Twining went on to become chairman of the Joint Chiefs of Staff.) The Thirteenth originally played a defensive role, standing guard over the strategic islands of the South Pacific. Later, the command swung to the offense. It traveled an air-road northwest from the Solomons to the Admiralty Islands, New Guinea, Moroai and the Philippines. This island hopping organization did not fight from centralized bases, but often used grass runways captured from the Japanese only hours before. During this period the command earned the nickname "Jungle Air Force." Its B-24 Liberators flew missions as far as 3,000 miles and as long as 18 hours. From January 1944 until V-J Day, Jungle Air Force units were spread over 40 islands and the continents of Australia and Asia. Pilots of the Thirteenth originated many low level attack techniques still in use today. Aircraft of the Thirteenth flew over 12 seas and participated in 13 campaigns.

In the 1970s, the Thirteenth Air Force operated from three countries — the Republic of the Philippines, Thailand, and the Republic of China (Taiwan) — that's an area of about one-eighth of the world's surface, approximately 25 million square miles. In that area 250 million people speak more than 100 different dialects and practice six of the world's major religions.

388th Tactical Fighter Wing

Organized at Korat RTAFB in April 1966, the wing carried on a tradition of combat excellence dating back to the decisive strategic bombing campaign of World War II in Europe. When it was organized at Korat Royal Thai Air Force Base, the 388th Tactical Fighter Wing inherited the history and honors of the 388th Bombardment Group (Heavy), which had been activated December 24, 1942, at Gowen Field, Idaho.

In June 1943 the group deployed to England, where it remained throughout the war as one of the key units of the Eighth Air Force. During World War II, the unit earned two

This view looking north shows Korat, including its runway, parking ramp and base facilities.

SOUTHEAST ASIA

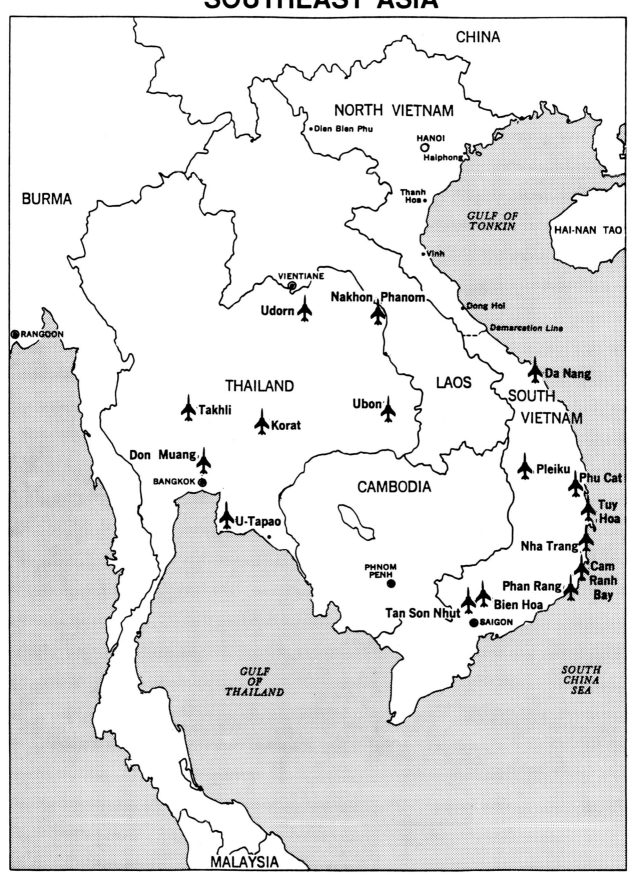

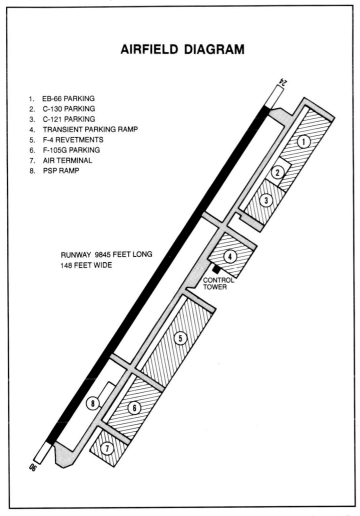

AIRFIELD DIAGRAM

1. EB-66 PARKING
2. C-130 PARKING
3. C-121 PARKING
4. TRANSIENT PARKING RAMP
5. F-4 REVETMENTS
6. F-105G PARKING
7. AIR TERMINAL
8. PSP RAMP

RUNWAY 9845 FEET LONG
148 FEET WIDE

CONTROL TOWER

Distinguished Unit Citations while hitting a variety of significant targets.

In August 1945 the organization returned to the United States and was inactivated. It returned to the active rolls March 23, 1953, as the 388th Fighter-Day Wing, and in late 1954 moved to Etain, France, during the buildup of NATO forces. Almost three years later the unit was inactivated in France, but it was reactivated as the 388th Tactical Fighter Wing at McConnell Air Force Base, Kansas, October 1, 1962. There, the wing flew F-100 and F-105 fighters.

Inactivated February 8, 1964, the 388th Tactical Fighter Wing returned to the active list March 14, 1966. It then organized at Korat Royal Thai Air Force Base 25 days later, absorbing the personnel, equipment, and resources of the 6234th Tactical Fighter Wing, which had operated at Korat for a year.

Though assigned to the Thirteenth Air Force, during the Vietnam conflict the 388th TFW traced its command structure back through the Seven/Thirteenth Air Force commander at Udorn RTAFB. The Seven/Thirteenth coordinated the air war in Southeast Asia using the combined air power of the Thirteenth Air Force wings in Thailand and the Seventh Air Force wings in South Vietnam.

At the time of its reactivation the 388th Tactical Fighter Wing was assigned two tactical fighter squadrons flying the Republic F-105 Thunderchief — the 469th and 421st. As part of Rolling Thunder, the first major bombing campaign against North Vietnam, Thunderchiefs from the 388th TFW conducted daily strikes against North Vietnam from April 8, 1966, until the bombing restrictions imposed March 31, 1968. During this time aircrews of the 388th dropped more than 90,000 tons of ordnance in raids on all major targets in North Vietnam. From March 31 to November 1, 1968, the Thunderchiefs confined their activity to the southern panhandle of North Vietnam. Major missions flown by the 388th Tactical Fighter Wing during its first two years included the first (March 10, 1967) and subsequent strikes on the Thai Nguyen iron and steel works and thermal power plant; raids on all major communist airfields (including the first against Phuc Yen MiG base, October 24, 1967 and the first (August 11, 1967) and subsequent attacks on the Hanoi (Paul Doumer) highway and railroad bridge crossing the Red River in downtown Hanoi. The last and most effective Rolling Thunder raid on the bridge took place December 18, 1967, destroying eight and damaging 19 spans to leave it completely unusable. In air-to-air encounters, a 388th Tactical Fighter Wing pilot (Major Fred L. Tracy, June 29, 1967) flew the first F-105 to shoot down a Communist MiG-17. Three times the 388th Tactical Fighter Wing has been honored by being named a recipient of the Air Force Outstanding Unit Award.

In 1972 the 469th Tactical Fighter Squadron, the oldest tactical fighter squadron flying in the Southeast Asian conflict from Thailand, flew its 60,000th combat mission in March 1970. This was believed to be a Thailand record.

At the beginning of 1972, the 469th was still part of the wing along with its sister squadron, the 34th. Both squadrons had received their F-4E Phantom IIs from the 33rd TFW at Eglin AFB, Florida during 1969. The 6010th Wild Weasel Squadron (WWS), formed in November 1970 and redesignated the 17th WWS on December 1, 1971, was also stationed at Korat flying F-105Gs. The 42nd Tactical Electronic Warfare Squadron (TEWS) (being renumbered from the 6460th TEWS in January of 1968) flying EB-66s was the other active flying unit of the 388th TFW. In early April 1972 the 561st TFS from McConnell AFB deployed to Korat with their F-105G as part of Constant Guard I. In June of 1972 the 35th TFS, flying F-4Ds arrived from Da Nang and was temporarily assigned to the 388th TFW. The 35th TFS had left the 3rd TFW at Kunsan AB, Korea in April and, after a stay of a little over two months with the 366th TFW at Da Nang, came to Korat. It operated at Korat until October 10, 1972.

As part of the build up in support of LINEBACKER operations seven tankers and twelve crews were assigned to Korat as the 4104th Air Refueling Squadron (Provisional) under the nickname Tiger Claw. The aircraft arrived between June 12th and 17th and remained until October 10, when they transferred to U-Tapao, making more ramp space available for the incoming A-7s from the 354th TFW.

This four ship of 469th TFS F-4Es stopped in the de-arm area. After returning from a mission the aircraft were required to stop in the de-arm area to have safeing pins removed before takeoff re-installed before taxiing back to their parking spots.

On October 10, 1972, 72 A-7Ds of the 354th Tactical Fighter Wing arrived from Myrtle Beach, South Carolina and replaced the F-4s of the 469th TFS and the 35th TFS. The 354th TFW A-7s remained at Korat until April 5, 1974. Some of the aircraft were reassigned to the 3rd TFS when it was activated as part of the 388th TFW on March 15, 1973. The 3rd TFS remained at Korat until December 15, 1975.

On November 28, 1975, the 388th TFW was relieved of its operational commitments and began preparing to transfer all its units back to the United States. By December 15, 1975,

all the Wing's aircraft had departed Korat and on December 23, the Wing formally departed Korat for its new home at Hill AFB, Utah.

Presently the 388th Fighter Wing flies F-16C/Ds and is based at Hill AFB, Utah. The 34th TFS is still part of the 388th. The 421st Fighter Squadron, once part of the 388th at Korat then equipped with F-105s, is also presently an F-16 squadron assigned to today's 388th. The Wing's third F-16 squadron is the 4th Fighter Squadron.

KORAT AIRCRAFT

F-4E Phantom II

The F-4Es assigned to the two fighter squadrons at Korat during 1972 were mostly 1967 models, with a few 1966, 1968 and two 1969 model aircraft. The F-4Es were powered by General Electric J79-GE-17A/F engines, and differed from the other F-4s operated by the U.S. having an internally

mounted M61A1 20mm six barreled Gatling gun and the AN/APQ-120 fire control radar.

During 1972, while still flying combat missions, the Korat F-4Es were in the process of being modified. The Korat F-4Es initially did not have the extended gun fairing. The new

The 20mm M61A1 nose gun is seen here with the fairings removed.

In this photo of 68-322 the ECM pod can be seen mounted in the forward Sparrow (AIM-7) well.

69-7551, one of the 469th TFS newest (a block 44 aircraft), with the sharkmouth removed, heads for a strike in northern Laos.

67-392, flying here with 67-298 was credited with downing a MiG-19 on September 3, 1972 using an AIM-7 Sparrow missile.

66-313 returning from a bombing mission with its bomb racks empty.

67-315 and 67-385, as evidenced by the extended air refueling receptacle on 315, approach precontact position for refueling from a KC-135A. These two aircraft are flying as part of a 469th TFS four ship. The two F-4 squadrons would normally share aircraft and the crews flew which ever squadron's aircraft were assigned to the mission.

longer fairing, designed to shield the flash from the gun barrels, was added to Korat's F-4s during late January and early February of 1972. The inboard pylons on Korat's F-4s had not yet been modified to allow carriage of the AIM-9 missile rails on the side of the pylon. As a result, when carrying AIM-9s, the dual missile rails were attached to the ejector rack and hung from the bottom of the pylon. The aircraft were all "slick wing" aircraft, that is, they had not yet had the leading edge maneuvering slats added. They did not have chaff dispensers installed, and as a result when chaff was required it was placed in the speedbrakes. If the speedbrakes were deployed, the chaff fell out and was dispersed by the airflow under the wing. Most aircraft had the forward AIM-7 wells modified to allow carriage of electronic countermeasure (ECM) Pods. Carrying the pods in the forward missile wells kept the inboard wing pylons free to carry additional bombs.

Korat F-4Es carried ECM jamming and deception pods which were designed to jam the SA-2s frequency band and inhibit the SA-2 SAM's ability to track the aircraft. F-4 flights were grouped into two elements of two aircraft each, flying "pod" formation, the overlapping protection gained from the pods provided satisfactory results.

ECM pods were controlled by the WSO from the rear cockpit. They were used to protect the aircraft from SAMs by jamming Fansong fire control radars. The Fansong radar detected and tracked the target aircraft, and at the same time guided the SA-2 SAM (Guideline missile) to the target aircraft. The pods defeated the SAM system in one of two ways; either by blanking the radar scope, preventing the radar operator from finding the target aircraft, or by putting false targets on the radar making target identification by the operator more difficult. The F-4Es had radar warning gear, also controlled from the rear cockpit, which notified the aircrew when a radar was transmitting. It further identified the radar type and in some cases the mode (target track, launch, guidance, etc.) in which the radar was operating. It displayed a strobe which indicated the direction to the radar whose signals were being received.

The 34th Tactical Fighter Squadron (JJ tail code) and 469th Tactical Fighter Squadron (JV tail code) were the 388th's two fighter squadrons during 1972. The 469th TFS was disbanded in October 1972. The 34th TFS used black as their squadron color and the 469th used bright green. All the F-4Es of the 388th TFW had shark mouths painted on the nose and gun fairing below the radome.

67-269 banks away showing its load of 12 MK82 500 pound bombs. The bombs on the wing pylons are fitted with fuze extenders. Noteworthy is the lack of AIM-7 Sparrow missiles. Two Sparrows were carried anytime the mission took the F-4 within striking range of the North Vietnamese MiGs.

The ECM pod mounted in the right forward missile well seen here on 67-287 was used to combat the SA-2 surface to air missile (SAM) system.

67-316 returning from a January 1972 bombing mission.

This bottom view of 67-315 shows the two AIM-7E missiles in the aft Sparrow wells and empty bomb racks. A strike camera is installed in the right forward Sparrow well.

67-284, in close formation, is returning to Korat after a bombing mission.

66-313, seen here during a formation landing approach, was Korat's other 313. 66-313 had a non-standard tail number with the year digits (66) in white rather than black.

67-269, seen here, is carrying its left Sparrow missile in the forward well. The Sparrows were always carried in the aft wells, unless the aft missile launcher equipment was not operational. Using the aft missile wells left the forward wells available for ECM pods or strike recording cameras.

69-7267 early morning low level over Thailand on its return to Korat following night escorting an AC-130 gunship.

This C-5 is unloading external fuel tanks to replace those jettisoned over North Vietnam.

Above and below: C-141s provided the scheduled airlift service into Korat, carrying both personnel and materiel.

69-7267 low level over Thailand. Korat's F-4s normally flew very little at low level, staying at an altitude of at least 6000 feet in order to keep out of the range of small arms fire.

68-313, assigned to the 34th TFS, taxis out for a strike mission during January 1972.

67-311, assigned to the 469th TFS, taxis out for a strike mission during January 1972.

Cargo C-130Es regularly stopped at Korat delivering aircraft supplies and replacement parts.

69-7551 claimed a MiG-21 kill when flying as a 34th TFS aircraft. The star painted on the intake splitter remained after the claim was denied. (Don Logan Collection)

69-7551 photographed in early February 1972. The aircraft carries a COMBAT SAGE firebee kill marking on the intake.

69-7551 undergoing maintenance in its revetment. Due to the fact that Korat had very limited hangar space, most maintenance was accomplished outside.

67-269 carrying an unusual load with a AIM-7 in the left forward missile well and an ECM pod in the right forward well.

67-385 flying a wing approach for a landing at Korat.

An engine dolly sits off the right wing of 67-274 on the ramp at Korat.

67-296 loaded with CBUs which were the primary weapon used against SAM and AAA sites.

67-309 in a first row revetment loaded with MK82s for a strike mission.

Flying level at 17,000 feet on a LORAN delivery, 66-313 releases its ordnance at "pickle" (bomb release) command of the lead aircraft.

The 34th TFS and 469th TFS, the 17 WWS, and the 42 TEWS put crewmember's names on their aircraft. As seen here on my aircraft (69-7551) the background was the squadron color with the names in white.

Above: 69-7551 was "my aircraft." Crewmembers normally did not fly the aircraft with their name on it more frequently than other aircraft of the type in the wing. Though assigned to the 469th TFS, many of my missions were flown in 34th TFS aircraft.

Right: Due to limited hangar facilities, many maintenance functions including changing engines, were accomplished on the ramp. Shown here is a J79-GE-17A/F, on its handling trailer, ready to be installed in a 469th TFS F-4E.

If Korat aircraft had to divert in to other bases, they would sometimes receive "ZAPs" from the units who supported them. This Triple Nickel ZAP was painted by 555th TFS personnel after a diversion into Udorn RTAFB.

This ZAP was applied by personnel of the 433rd TFS at Ubon RTAFB.

This aircraft illustrates a different version of the 555th TFS ZAP.

F-4D Phantom II

F-4Ds from the 35th Tactical Fighter Squadron (UP tail code, light blue tail caps), 3rd Tactical Fighter Wing, Kunsan AB, Republic of Korea were assigned to Korat from June 6, 1972 through October 10, 1972 to augment the F-4Es during LINE-BACKER missions. After the inactivation of the 469th TFS, some F-4Ds were assigned to the 34th TFS.

A four ship of 35th TFS aircraft lines up on Korat's runway for takeoff. The aircraft are carrying CBUs on their outboard wing pylons, ECM pods on their inboard pylons, and fuel tanks on centerline.

65-691 an F-4D from the 35th TFS, 3rd TFW sits on the temporary ramp at Korat. (Jerry Geer Collection)

F-4Ds from the 35th TFS, 3rd TFW deployed to Korat to augment Korat's F-4s during LINEBACKER missions. (Jerry Geer Collection)

This F-4D (65-691) of the 356th TFS transferred to Korat as one of the 35th TFS aircraft, arriving in June of 1972 as part of Constant Guard I.

This 469th TFS F-4E, 67-288 with the nickname "LAURA-LU" on the nose gear door was one of the few Korat aircraft to be given a nickname.

Korat had two F-4Es with 313 as the last three digits of its serial number. 68-313, seen here, was assigned to the 34th TFS and 66-313 assigned to the 469th TFS.

A-7D Corsair II

A-7Ds of the 354th Tactical Fighter Wing from Myrtle Beach AFB, South Carolina operated out of Korat from October 16, 1972 through May of 1974. The aircraft deployed to Korat as part of Constant Guard VI in support of the stepped up bombing of North Vietnam during the LINEBACKER campaigns. Initially, the entire 354th TFW, carrying MB tail codes deployed to Korat. The 354th TFW Tactical Fighter Squadrons at Korat included the 353rd with red as the squadron color, 355th (blue), and 356th (green). While the A-7 remained at Korat, the squadrons periodically rotated to and from the United States. A-7 crews from Davis Monthan AFB also saw action flying from Korat. As the A-7s arrived, they replaced the F-4Es of the 469th TFS and the F-4Ds of the 35th TFS.

The A-7Ds flying from Korat also replaced the A-1s from Nakhon Phanom RTAFB as the "SANDY" close air support and helicopter escort aircraft used in Search and Rescue missions. This resulted from the transfer in October of 1972 of the A-1s to the South Vietnamese Air Force. The A-7s

The A-7s took over the taxi through revetments vacated by the departing F-4s. (Don Logan Collection)

proved themselves in this job, taking part in some of the most dramatic rescue efforts of the Vietnam war.

The 388th TFW gained its own A-7D unit in March of 1973 with the activation of the 3rd Tactical Fighter Squadron. A-7s of the 3rd TFS carried JH tail codes.

70-982 low level returning to Korat after a combat mission in SEA. (Don Logan Collection)

Posed in front of the "Welcome To Korat" sign, shortly after arriving at Korat is this A-7D (70-957) of the 356 TFS, 354 TFW. This photo was taken on October 26, 1972. (Don Larsen Via Pat Martin)

This front view of a 354 TFW A-7D shows MK82 bombs on the middle wing pylons. (Don Logan Collection)

Tactical fighter aircraft were required to stop in the arming area to have the "Remove Before Flight" safeing pins removed from their weapons and racks, and to have the last chance inspection completed on the aircraft. The pilot of this A-7D waits as the ground crew approaches from the left side of the aircraft. (Don Logan Collection)

70-934, in its revetment loaded for a combat mission. (Don Logan Collection)

WILD WEASEL AIRCRAFT

The Wild Weasel Program began in August of 1965. The purpose of the Wild Weasel aircraft is to locate radar systems associated with enemy air defenses and destroy them before they can destroy friendly aircraft. In September, North American Aircraft began converting the first two of an eventual seven Wild Weasel I F-100Fs. Initially the Wild Weasel equipment consisted of an E/F band panoramic receiver and a E-I band Vector VI radar warning receiver. During mid-November a C band launch warning receiver was installed in the four F-100Fs already converted. In late November the four Wild Weasel F-100s deployed to Korat for a 90 day Southeast Asia (SEA) operational test. After completing the test, in which one of the Weasels was lost to AAA, a second increment of Weasel aircrews and two more aircraft were deployed to Korat. The seventh F-100 Wild Weasel remained at Eglin AFB Florida for crew training until March of 1966, when it joined the other F-100 Weasels at Korat. The last F-100 Wild Weasel mission was flown in July of 1966. Of the seven aircraft, two were lost in combat (December 20, 1965 and March 23, 1966) and one was lost in an operational accident (March 13, 1966).

In October of 1965, two additional Wild Weasel projects (IA and II) entered testing. The Wild Weasel I project used a single seat F-105D as the airframe to be modified, while the Wild Weasel II used the two seat F-105F. In testing, the IA configuration proved that a single seat aircraft could not adequately perform the Weasel Mission. The Wild Weasel II test was a failure with the test never being completed. Based on the success of the Wild Weasel I F-100Fs, the Air Force directed Sacramento Air Materiel Area at McClellan AFB, California in December 1965, to install the F-100F Weasel gear in seven F-105Fs. This number was later increased to thirteen. Also Ogden Air Material Area at Hill AFB was di-

63-8266, with the standard missile load of an AGM-45 on each outboard and an AGM-78 on one inboard pylon, taxis out for takeoff.

rected to conduct a feasibility study of installing the same equipment in an F-4C.

Work on the Wild Weasel III F-105F began at McClellan in January of 1966, and before the first F-105F Weasels deployed to SEA, a new piece of equipment, an Azimuth-Elevation pointer, was added to the aircraft. This Az-El system projected a homing dot on the pilot's optical sight helping the pilot visually acquire the target.

Over the next few years the Air Force modified 86 F-105F to the Wild Weasel III configuration. Major and minor improvements were installed in the aircraft as they became available and local modifications were accomplished on some aircraft. As a result of these uncontrolled modifications there was no standard configuration for the Wild Weasel version of F-105F. The F-105G was created in an attempt to standardize the F-105 Weasel fleet.

This lineup of F-105Gs on the Korat flightline in April of 1972 still are marked with the ZB tail code. The 17th WWS tail code was changed to JB in June of 1972.

This bottom view of a Korat F-105G Wild Weasel shows the standard SAM suppression load of a AGM-45 Shrike on each outboard wing pylon, an AGM-78 Standard ARM on the right inboard wing pylon, and fuel tanks centerline and left inboard pylon. The F-105Gs also carried an auxiliary fuel tank in the bomb bay.

63-8292 seen over the southern province of North Vietnam on a SAM suppression mission.

62-4423 approaches the boom for a pre-strike refueling.

F-105G Wild Weasel

F-105G Wild Weasel aircraft were two seat, single engine (Pratt & Whitney J57-P-19W afterburning turbojet) aircraft. The G model had been made by converting F-105F two seat trainer versions of F-105D fighter/bomber to the Weasel configuration by adding the electronic systems necessary for their Wild Weasel role. Like the F-4E, the F-105G had an internally mounted M61 20mm gun. The F-105s, differing from any other aircraft then operating in Southeast Asia, had both a receptacle and a probe type inflight refueling system. The two seat version of the F-105 was 30.5 inches longer forward of the wing. To compensate for the added forward area, the area of the vertical fin was increased by 15%.

By 1972 the Wild Weasel F-105G had the following capabilities and Weasel equipment:

• Launch capability for Standard ARM Mod 1 (AGM-78B/C)
• Launch capability for the AGM-45 Shrike
• APR-35 Radar Receiving Set (Panoramic/Analysis Receiver)
• ALR-31 Countermeasures Receiving Set
• APR-36 Radar Receiving Set (Radar Warning Receiver)
• APR-37 Radar Receiving Set (Missile Launch Warning)
• QRC-373 ECM Set (Search/Acquisition Radar Jammer)
• ALQ-105 ECM Set (Jammer contained in the long fuselage blisters under the wings)
• 14 Channel Recorder

F-105 Weasel Operations

The 17th Wild Weasel Squadron (originally the 6010th WWS) flew F-105G Wild Weasel aircraft while operating as part of the 388th TFW. The F-105Gs carried a ZB tail code during the first half of 1972. The tail code was changed to JB during June of 1972. The squadron color was medium blue. The 17th WWS was supplemented during LINEBACKER and II with aircraft and crews from the 561st Tactical Fighter Squadron deployed from McConnell AFB, Kansas.

On April 6, 1972, as Detachment A of the 561st TFS, a majority of the 561st TFS deployed to Korat for combat operations under a program called Constant Guard I. At Korat, the detachment operated under control of the 388th TFW. Most of the 561st's F-105Gs, crews, and a large number of maintenance personnel operated at Korat from April 8 to early September, and a smaller force continued under Constant Guard combat operations from Korat until January 27, 1973.

The aircraft from the 561st carried an MD tail code, with a fin cap painted in the squadron color of yellow. The 561st later, while still at Korat, carried a WW (Wild Weasel) tail code. This tail code has remained with the squadron, and was carried on 561st F-4Gs stationed at George AFB until June, 1992.

From June 15, 1973 to September 7, 1973, Detachment A continued operations at Korat, with 12 F-105G's and 180 maintenance personnel. The aircraft and personnel returned to George AFB during the first week of September 1973.

The F-105Gs remaining after the end of the Vietnam war were reassigned to the 35 TFW at George AFB California, and then in the summer of 1979 were reassigned to the 128th Tactical Fighter Squadron, 116th Tactical Fighter Wing of the Georgia Air National Guard at Dobbins AFB. The F-105Gs were retired and began leaving operational service in October of 1982.

62-4423 in parked in the open on the Korat flightline.

63-8300 taxis for takeoff with a full missile load.

63-8301 in the arming area before takeoff.

63-8333 was shot down on February 17, 1972. Both crew members were captured and held as POWs.

62-4423 and 62-4416 on tanker.

63-8292 in its parking area loaded with an AGM-45 Shrike missile.

62-4423 taxis out for takeoff for a SAM suppression mission.

Two Korat F-4s, 67-290 and 67-316, are holding on the right wing of a KC-135A awaiting their turn on the boom. (USAF)

The 469th TFS and the 34th TFS had a group of five pilots who made up the "Thunder Buzzards." At the end of a squadron members last scheduled mission, his End Of Tour (EOT) flight, the Thunder Buzzards, on their 90cc Hondas trailing red and green smoke, would escort his aircraft back to its parking spot.

67-379 taking on fuel from KC-135A 58-0078.

63-80305 here flying in close wing tip formation with a Korat F-4E.

This 561st TFS F-105G (63-8342) was shot down on April 15, 1972 shortly after its arrival. It was shot down by a SAM, with both crewmembers lost, listed as MIAs.

The flight line at Korat in April 1972 after the arrival of the 561st TFS.

Contrary to the appearance of this photo, F-4E 67-379 brakes down and away from the KC-135A as the tanker continues its post refueling straight and level flight.

F-4E 66-322 takes on fuel from KC-135A 56-3608.

34th TFS F-4E 67-315 takes on fuel from KC-135A 59-1479 while 469th TFS F-4E 67-385 waits its turn.

F-4C Wild Weasels

McDonnell Aircraft began the initial Wild Weasel IV F-4C modifications in April of 1966. The F-4C Weasel had most of the same equipment as the F-105F Weasel, but lacked the capability to carry and launch the AGM-78 Standard Arm. They did not receive the upgrades which made an F-105G out of the F-105F, so by the time they deployed to Korat, they were no longer equivalent to the F-105Gs — they were a generation behind them.

Six Wild Weasel F-4Cs from the 67th TFS (ZG tail codes) 18th TFW Kadena AB, Okinawa deployed to Korat to augment the F-105Gs and operated in support of LINEBACKER operations between September 9, 1972 and February 18, 1973. This deployment was the combat debut of the Wild Weasel F-4Cs.

The F-105G Wild Weasels of 17th WWS and the 561st TFS were augmented by the less capable F-4C Wild Weasels like this one (63-7474) of the 67th TFS photographed at Korat on November 1, 1972. (Don Larsen Photo Via Pat Martin)

ECM/JAMMING AIRCRAFT

EB-66C and EB-66E

The EB-66C and EB-66E were Electronic Countermeasures versions of the Douglas B-66 Destroyer. Though externally appearing quite similar to the Douglas A-3 Skywarrior, they were very different. The B-66s were powered by two Allison J71-A-13 turbojet engines which were much less reliable than the Pratt & Whitney J57-P-10 of the A-3. The B-66s did not have the folding wings and tail of the A-3. B-66s did however have ejection seats, a feature lacking in the A-3. EB-66s were inflight refuelable through a six foot long probe attached to the right side of the nose. The EB-66 aircraft were used as a radar systems and communications jamming platform with the assigned mission of blocking the North Vietnamese communications and early warning network and thereby neutralizing their defensive system.

The EB-66C version started out as an ELINT (Electronic Intelligence) version of the EB-66 with only a few jammers for self protection. By 1972, after several equipment upgrades the C had a substantial jamming capability, though not equivalent to the EB-66E in terms of the number of jammers. The C model could perform "smarter" jamming because the C had four additional Electronic Warfare Officers as crewmembers riding in what had been the bomb bay. They could analyze threats and direct the steerable jammers to selectively jam the threats. The EWOs were seated in downward firing ejection seats, arranged in two forward facing rows. Both models had tailcone-mounted IR countermeasures and chaff dispensers. The EB-66C could be distinguished from the EB-66E by the two antenna radomes with fairings (canoes) on the bottom of the fuselage, and the wing tip antenna pods. The EB-66E had only the three crew stations in the forward fuselage, the pilot, navigator and a single EWO. Though the E had over 20 jammers, they were not steerable and could not be directed at specific threats.

EB-66 Operations

The 42nd Tactical Electronic Warfare Squadron (TEWS) flew EB-66C and EB-66E aircraft. The aircraft carried a JW tail code with maroon as the squadron color. The 42nd TEWS was supplemented by eight additional EB-66s from the 39th TEWS based at Shaw AFB. These aircraft arrived on April 11, 1972 under the Constant Guard I deployment.

At the end of 1973, the remaining EB-66s were retired. One aircraft was salvaged in place at Korat, and the other were flown, between January 2 and 17th, to Clark AB, The Philippines where they were scrapped.

Above: EB-66E 54-522 on takeoff at Korat. Below: EB-66E 54-540, without a tail code, sits in its revetment at Korat.

EB-66E 54-443 sits in its parking spot as a 34 TFS F-4E crosses over the runway threshold. (Vic Hilgren)

This EB-66 can be identified as a "C" model by the two antenna cones on the bottom of the aircraft. The air refueling probe visible on the nose was painted in a barber pole pattern of white and red. (USAF)

This EB-66 can be identified as a "E" model by the multiple antennas on the bottom of the aircraft. (USAF)

EB-66C 55-385 prepared for a combat mission awaits the aircrew. The power carts which provide both electrical power and compressed air for engine start are visible.

EB-66C 54-468 taxis to runway 06 for a mission over North Vietnam.

54-466 (later shot down as BAT-21) awaits its crew for a combat mission

EB-66C 54-479 refuels from KC-135A 58-032. (USAF)

Above and below: 54-520 and 54-479 on the engine run pads at the far end of the EB-66 ramp.

This EB-66E, 54-509, banking away from the photographer, displays its numerous belly antennas and shows the similarity of the South East Asia camouflage to the terrain below. (Vic Hilgren)

54-466, shot down on April 5, 1972, is seen here taking on fuel from a drogue equipped KC-135A. (Vic Hilgren)

EB-66E 53-479 taxis out for takeoff at Korat. (Vic Hilgren)

In this front view of an EB-66E, the B-66 similarity to the Navy A-3 is readily visible.

In keeping with other 388th wing aircraft this EB-66E also has a sharkmouth.

The crew boarding ladder on this EB-66E (54-479) is visible behind the nose gear.

This aircraft, 54-389 can be identified as an EB-66C by its wing tip antennas and the canoe radomes on the bottom of the fuselage.

EB-66E (54-520) sits in the arming area awaiting takeoff. The EB-66's main landing gear retracts to the rear and up into the fuselage.

EC-121 "DISCO" AWACS

The EC-121 was a military version of the Lockheed L-749 Super Constellation and were powered by four Wright R-3350-91 Turbo-compound radial piston engines. The EC-121 carried the Airborne Surveillance and Control System (ASACS) developed as part of the COLLEGE EYE program.

The EC-121 AEW&CS aircraft belonged to the 552nd Airborne Early Warning and Control Wing (AEW&CW) as-signed to McClellan AFB in Sacramento, California. A de-tachment of the 552nd AEW&CW was stationed at Korat. These aircraft, using the call sign DISCO, monitored North Vietnamese aircraft movement and communication, and di-rected the U.S. MiGCAP flight intercepts. The EC-121s were retired in 1976 when the 552nd AEW&CW started to receive the new Boeing E-3A AWACS.

This EC-121 133 of the 552nd AEW&CW, call sign DISCO, taxis out for takeoff. The COLLEGE EYE EC-121s did not have the separate height finding antenna and therefore did not have the large hump radome carried on earlier EC-121 versions.

EC-121 133 accelerates down runway 06 for takeoff. Korat's runway had an arresting cable 800 feet from each end. The location of the cable was marked by a black sign with a yellow ball painted on it, visible behind the 121. The checkerboard building in the background holds the cable brake and the machinery required to rewind the cable.

This EC-121 (52-3418) lifts off for a DISCO mission monitoring the airspace over North Vietnam.

52-3418 taxis for takeoff.

The wide radome under the wing and fuselage contains the radar antenna used surveillance of the North Vietnamese air space.

Above and below: The Navy equivalent to the Air Force's DISCO aircraft like this EC-121M from VQ-1 would sometimes stop at Korat.

52-3412 in its revetment at Korat.

EC-121R "BAT CATS"

The EC-121Rs stationed at Korat and flown by the 553rd Reconnaissance Wing "Bat Cats" were also based on the Lockheed L-749 Super Constellation and were used as radio relay aircraft for the sensors "seeded" on the Ho Chi Minh trail.

A BAT CAT EC-121R sits in its revetment at Korat.

The EC-121Rs of the 553rd Reconnaissance Squadron, call sign BAT CAT, were camouflaged in the standard Southeast Asia colors.

This EC-121R 29473 is parked at the end of the line of revetments at Korat. A C-141 in natural metal can be seen in the background parked in the another revetment across the ramp.

WC-130s often stopped at Korat as part of their weather surveillance missions in Southeast Asia.

C-7A Caribous, like this one from the 457th Tactical Airlift Squadron of the 483rd Tactical Airlift Wing at Cam Ranh Bay South Vietnam, made frequent visits to Korat.

Airlift service between the bases within Thailand was supplied by C-47s like this one.

KC-135A STRATOTANKER

The 4104th Air Refueling Squadron, (Provisional), attached to the 310th Strategic Wing Provisional based at U-Tapao, was activated with seven aircraft and twelve crews at Korat on June 12, 1972, The squadron remained in place flying six sorties per day until it was inactivated on October 10, 1972. Its aircraft were then transferred to U-Tapao to make room for the A-7Ds of the 354th TFW deploying to Korat. The KC-135A is the tanker version of the Boeing C-135 aircraft. The aircraft is powered by four J-57 engines with water injection thrust augmentation. The KC-135 has a crew of four, a pilot, copilot, navigator, and a boom operator.

KC-135A (57-1460) assigned to the 11th Air Refueling Squadron (ARS) at Altus AFB, Oklahoma, seen here temporarily assigned to the 4104 ARS at Korat. As evidenced here, the Korat Sharkmouth was even applied to tankers. (USAF via Joe Bruch)

In this view from the boom pod, 67-0316, with the author in the back seat, leaves the tanker after receiving its fuel. (USAF)

An EC-130E ABCCC at rotation on takeoff at Korat.

C-130 VARIANTS

C-130E ABCCC

A detachment of the 7th Airborne Command and Control Squadron, operating C-130E ABCCC (Airborne Battlefield Command and Control Center) aircraft, was assigned to Korat on April 30, 1972. These aircraft were C-130Es which had been modified to allow carriage of the ABCCC AN/USC-15 Capsule. The capsule had a battlestaff crew of 12, with numerous HF, UHF, VHF, and FM radios to allow communication with both ground and air forces. A JC tail code was added to the aircraft in mid-1973.

Above and below: The EC-130E ABCCC aircraft can be identified by their avionics cooling scoops on both sides of the fuselage in front of the wing root and their communications antennas pointing forward from under the wing outboard of the engines.

68-313 closes into refueling position to receive fuel from 60-0349.

Though not based at Korat, WC-130s like this one on Korat's PSP parking ramp were frequent visitors.

HC-130P "KING" Search and Rescue

The 56th Air Rescue and Recovery Squadron operated HC-130P Search and Rescue (SAR) aircraft from Korat during the last three quarters of 1972. These aircraft were modified HC-130Hs with added equipment giving them the capability to refuel helicopters in flight.

Left: In this front view of an HC-130P the radome for the search radar is visible on the top of the fuselage. The outboard wing pylons contain the hose and basket used for inflight refueling of helicopters.

Above and below: HC-130Ps like these in the revetments at Korat were use as both the on scene command center for rescue efforts and the tanker for Jolly Green Giant helicopters. These aircraft also have the booms on the nose used as part of the Fulton recovery system.

THE AIR WAR - 1972

Before the resumption of the bombing of North Vietnam in April of 1972, the 388th's primary mission involved the interdiction of supplies and troops moving south along the Ho Chi Minh trail into South Vietnam. Occasionally the F-4Es were tasked to support friendly forces battling the communists in Northern Laos. The F-105Gs and EB-66s were tasked to fly electronic reconnaissance missions around North Vietnam, updating the Electronic Order of Battle (EOB). The EOB was an intelligence package which identified the types, numbers, and locations of enemy radar and communication systems.

In response to a full scale invasion of South Vietnam which began on March 30, 1972, President Nixon decided to resume bombing of North Vietnam. Starting on April 6, 1972 units of the 388th TFW took part in FREEDOM TRAIN bombing operations across the Demilitarized Zone (DMZ) into North Vietnam, with a restriction not to bomb north of the 20th parallel. A few strikes were ordered north, above the 20th parallel, into the Hanoi/Haiphong area. The first occurred on April 16 and was called FREEDOM PORCH BRAVO. The strike was a coordinated plan for an intensified one day strike which combined both Tactical-Air and B-52 resources in a four hour attack of key North Vietnamese defenses and logistics targets. This strike was to become a forerunner of the future Tacair/B-52 operations which occurred during the first LINEBACKER campaigns (later referred to as LINEBACKER I) and during LINEBACKER II.

As a result of the increased bombing, the Tactical Air assets at Korat were increased steadily during 1972 in support of the LINEBACKER bombing operations over North Vietnam. On March 30, 1972, before the resumption of bombing in North Vietnam, the following tactical aircraft were assigned to Korat:

F-4E	(34 TFS and 469 TFS)	34 aircraft
F-105G	(17 WWS)	19 aircraft
EB-66	(42 TEWS)	12 aircraft
		65 aircraft total

By May 13, 1972, at the beginning of LINEBACKER operations, the number of tactical aircraft at Korat had increased to the following:

F-4E	(34 TFS and 469 TFS)	34 aircraft
F-105G	(17 WWS and 561 TFS)	31 aircraft
EB-66	(42 TEWS + 8 Aircraft from the 39 TEWS)	20 aircraft
		85 aircraft total

In December 1972 at the height of LINEBACKER II operations, Tactical Air aircraft assigned to Korat had increased as follows:

F-4E (34 TFS)	24 aircraft
F-4C (Wild Weasel) (67TFS)	6 aircraft
F-105G (17 WWS and 561 TFS)	23 aircraft
EB-66 (42 TEWS)	17 aircraft
A-7D (354 TFW)	72 aircraft
	142 aircraft total

LINEBACKER

The LINEBACKER campaign began on May 9, 1972. LINEBACKER was a coordinated campaign of air operations against North Vietnam designed to disrupt the enemy's war supporting resources, and for the first time in the Vietnam war, included mining the harbors and ports in North Vietnam. The orders were:

"Conduct a continuing Tacair interdiction effort, augmented by B-52 sorties as required, to destroy and disrupt enemy POL and transportation resources and LOCs in NVN, for example POL storage and pumping stations, rails and roads, bridges, railroad yards, heavy repair equipment, railroad rolling stock and trucks. Utilization of resources to neutralize defenses is also authorized."

B-52Ds like this one, augmented with B-52Gs, brought heavy strategic bombing to Hanoi during LINEBACKER II. (USAF)

This B-52D is carrying CBU on the underwing pylons. (Don Logan Collection)

This view of the bomb bay of a B-52D shows the full load of M-117 750 pound bombs. (Don Logan Collection)

B-52Ds carried 108 MK82 500 pound bombs, 24 under each wing shown here, and 60 in the bomb bay. (Don Logan Collection)

NORTH VIETNAM STRIKE MISSION

During LINEBACKER, for USAF strikes against targets in the Hanoi/Haiphong area, a typical force composition was made up of 85 to 107 aircraft including the following:

Strike Force (20-28 aircraft)

8-12   Strike F-4s with laser guided bombs (normally from Ubon RTAFB)
12-16   Strike F-4s with conventional ordnance (normally from Korat)

Support Force (65-79 aircraft)

4	F-4 Weather Reconnaissance aircraft
3	EB-66 ECM Jammers from Korat
8-12	F-4 Chaff Bombers
8	F-4 Chaff Escorts
32	MiGCAP F-4s (normally from Udorn RTAFB)
16-20	F-4 Strike Escorts
8-12	Iron Hand Wild Weasel F-105Gs from Korat
4	Barrier CAP F-4s
2	RF-4C Photo Reconnaissance aircraft from Udorn RTAFB
2	F-4 Photo-Reconnaissance Escorts

In addition each mission had its usual search and rescue (SAR) forces, AEW&CS EC-121s, and numerous KC-135As to keep the strike force fully fueled.

LINEBACKER II

LINEBACKER missions continued though the summer, fall and early winter of 1972, influencing the North Vietnamese to return to the negotiation table. After the North Vietnamese walked out of the peace negotiations on December 13, President Nixon, having just been re-elected to another four year term, partially on a platform of pulling out of Vietnam, decided to commence a maximum effort bombing campaign to force the North Vietnamese back to the negotiating table. On December 17th, President Nixon sent the following message to the Commander of Pacific Forces, Commander Strategic Air Command, and 7th Air Force operational commanders:

"You are directed to commence at approximately 1200Z, 18 December 1972 a three day maximum effort, repeat maximum effort of B-52/Tacair strikes in the Hanoi/Haiphong areas against the targets contained in (the authorized list). Object is maximum destruction of selected military targets in the vicinity of Hanoi/Haiphong. Be prepared to extend operations past three days, if directed."

The North Vietnamese air defenses facing the U.S. were formidable. There were 32 operational SA-2 sites and six confirmed SAM operating areas in North Vietnam. These sites provided defensive coverage in and around the Hanoi/Haiphong area as well as along the northwest and northeast rail lines. Approximately 2300 SAMs and 200 launchers were scattered within the Red River Valley area. During the bombing lull of October and November 1972, the North Vietnamese had replaced the SAM assets expended during LINEBACKER. Over 2000 SAMs had been photographed by reconnaissance arriving from China by railroad. The North Vietnamese had 145 MiGs, with the majority assigned to defend Hanoi, based at Phuc Yen, Gia Lam, Yen Bai, and Kep airfields.

On December 17 and 18 intense activity was taking place at Korat and the other U.S. tactical bases in Southeast Asia to prepare for the new air offensive. The maintenance units were busy bringing as much equipment as possible to an operationally ready status. The LINEBACKER II schedule called for four complete strike/support packages per day, one tactical strike during the day and three B-52 night attacks.

To support these attacks the B-52 assets in Southeast Asia during December of 1972 had increased to 206 aircraft; 99 B-52Gs and 53 B-52Ds at Andersen AFB Guam and 54 B-52Ds at Utapao RTAFB Thailand.

The tactical day strikes used the same tactics used during LINEBACKER. The B-52 strike support packages were tailored to minimize the problems associated with night missions and the long time periods required to support three waves of B-52s.

The bomber support packages consisted of:

8	F-4 Chaff Bombers
3	EB-66 ECM Jammers from Korat
8	F-105G/F-4C Wild Weasels (from Korat)
10	F-4 Escorts
10	MiGCAP F-4s

In addition 15 to 33 F-111 missions against preemptive or pinpoint targets were launched from Takhli RTAFB.

LINEBACKER II continued for 11 days, with the last strike occurring on December 30. LINEBACKER II had convinced the North Vietnamese government it was time to go back to the negotiating table. On January 23, 1973, the peace treaty was signed in Paris which was to take effect on January 28. The treaty provided for the release of all American and Allied POWs and effectively put an end to U.S. involvement in the Vietnam conflict.

MISSION TYPES

Korat F-4 Phantom II Missions

During 1972, prior to LINEBACKER operations, the F-4 squadrons of the 388th TFW flew all types of combat missions using the F-4E as a "jack of all trades", a task for which it was so well suited. This differed from the other two USAF Wings F-4s based in Thailand. The 8th TFW at Ubon RTAFB used F-4Ds and various "smart" systems, including LORAN and PAVE KNIFE laser designators for the delivery air to ground ordnance. Some of the 432nd TRW's F-4Ds (assigned to the 13th TFS and 555th TFS), based at Udorn RTAFB, had, in addition to LORAN bombing and navigation systems, special COMBAT TREE air intercept equipment installed. Using this equipment, the Udorn F-4Ds flew a majority of counter-air missions over North Vietnam during 1972. The 432nd TRW's RF-4Cs flew most of the USAF reconnaissance missions.

Before the renewed bombing of North Vietnam, the F-4Es of the 388th TFW normally flew missions in Laos. These missions included supply route interdiction, missions in support of insurgent (anti-communist) ground actions, fast forward air controller (Tiger FAC) missions, and AC-130 and B-52 escort/anti-aircraft suppression. During LINEBACKER and LINEBACKER II the F-4Es missions into North Vietnam. These missions included interdicting supply routes and railroads, bombing military targets, and performing escort and barrier combat air patrol missions. In addition SAM hunter killer missions were flown by the F-4Es and F-105Gs Wild Weasels of the 388th TFW. Coincidentally starting in the mid-1970s some of Korat's F-4Es were converted to F-4G Wild Weasels, which replaced the F-105G in the Wild Weasel role. They saw combat in Desert Storm against Iraqi radar and SAM sites.

AIR TO GROUND MISSIONS

Air to ground missions included strike, interdiction, and close air support and rescue cover/escort missions. The weapons loads carried by the F-4E for air to ground were dependent on the target type. Flying out of Korat the F-4s, with their distinctive sharkmouths, carried one 270 gallon wing tank on each outboard wing station, leaving the inboard station on each wing and centerline available for ordnance. Normally two AIM-7E2 Sparrow missiles were carried in the aft Sparrow wells. The forward Sparrow wells carried ECM pods or a strike camera. The nose mounted M61 20mm Gatling gun was always loaded, but strafing was restricted by Wing regulations to SAR (rescue) missions or troops in contact missions only.

The primary visual weapon delivery mode used by Korat's F-4Es was dive toss. Dive toss mode was only available on the F-4Es and used the AN/APQ-120 fire control system radar to measure the range to the ground during a dive bomb run. In dive toss, the radar beam is boresighted, that is it is looked, aligned with pipper (bomb sight) on the pilots Lead

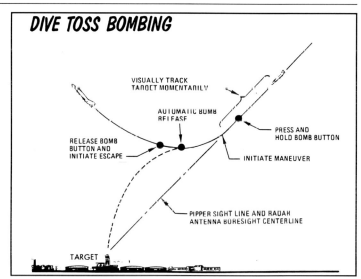

Computing Optical Sight System (LCOSS). As the pilot dives the aircraft at the ground, the WSO performs a radar lockon on the ground return. The pilot then flies the aircraft to place the pipper on the target. When the pipper is on the target, the pilot presses the bomb release (pickle) button. This marks the target for the AN/ASQ-91 Weapons Release Computer System (WRCS). He then begins the pull up away from the target. The WRCS uses the range from the radar along with the ballistics set in the WRCS by the WSO to compute a release point. This release is based on the location of the target (input by the pilot when the bomb release button was depressed). When the release point is reached, the WRCS sends a signal to release the weapons as selected by the pilot. Using dive toss allowed the F-4 to deliver weapons accurately while staying as much as 5,000 feet above the target. This kept the aircraft out of range of small arms (rifles, machine guns etc.) carried by soldiers on the ground.

When bad weather prevented visual delivery of ordnance, rather than returning to base with the unexpended ordnance, the F-4 flight normally rendezvoused with a Loran Pathfinder F-4 for weapon delivery using the Loran system. Loran equipped F-4s could be identified by the "Towel Rack" antenna on the top of the fuselage. The F-4 bomber flight flew close formation with the Loran Pathfinder, and on the Pathfinder's command, released the ordnance. The Pathfinder used the Loran equipment to navigate to and release ordnance against a known enemy area target. Because the target could not be seen, and the accuracy of this delivery left a lot to be desired, these missions were referred to by the aircrews as "TIC" (trees in contact, instead of troops in contact) missions or sometimes "Sky Pukes."

In poor weather conditions, if no Loran Pathfinder was available, the ordnance was delivered using direction from a ground radar site. This method of bomb delivery was called "Combat Sky Spot." The ground site directed the flight to the target area using radar vectors, and then cleared them to drop their ordnance. This method, like Loran, was inaccurate and was little more than a jettison of live ordnance in enemy territory

69-7551 taxis back, with racks empty from a combat mission.

67-379 and 69-7267 outbound on a bombing mission on December 25, 1971.

On October 6, 1972, 66-313 assisted 67-392 in downing a MiG-19 by maneuvering, causing it to fly into the ground.

RF-4C 68-604 of the 14th TRS, 432nd TRW at Udorn RTAFB is flying in pathfinder position for a LORAN directed release. The doors on the aft fuselage above the afterburner nozzles are for the Recce F-4s photo flash system. The flash cartridges are visible in the dispenser.

F-4D 66-8730 from the 497th TFS, 8 TFW from Ubon RTAFB leads a Korat F-4E fourship for a LORAN release. The 497th TFS flew most of their missions at night, and as a result, the undersides of their F-4Ds were painted flat black instead of the standard light gray.

67-392 is seen here in January of 1972 dropping 12 MK82s on a LORAN Pathfinder mission.

68-604 (OZ 14th TRS) leads the Korat flight in easy left turn to line up on the run in heading for a LORAN pathfinder release.

This 14 TRS, 432 TRW, RF-4C, (68-606), flying a lead as a LORAN pathfinder is carrying three fuel tanks and an ALQ-101(V)-1 on each inboard wing pylon.

Strike Missions

The majority of the strike missions flown by Korat's F-4s, in 1972, were against targets in North Vietnam. The purpose of these missions was to destroy military and industrial targets in North Vietnam supporting the North Vietnamese war effort. The normal ordnance carried on these strike missions was twelve Mk82 500 pound bombs, three on each inboard wing station and six on the centerline.

Two of my more memorable missions were typical of Korat strike missions. On 6 June 1972 I took part in a strike against the Kep airfield complex in North Vietnam. Our flight dropped ordnance on the runway resulting in numerous craters which put the airfield out of operation.

On 24 June 1972 I was flying as WSO in the number two aircraft of a four ship which was targeted against the Thai Nguyen Iron and Steel Works north of Hanoi. Our flight was responsible for heavily damaging the complex and causing a number of large secondary explosions.

67-208 seen here is loaded with 12 MK82 low drag bombs to be used against a target in North Vietnam.

67-385, is seen here in January 1972 with a load of 12 MK82s.

67-315, preflighted and loaded with 12 MK82s for its next combat mission, awaits the arrival of the aircrew.

The fuze extenders on the MK82 bombs can be seen protruding beneath the left wing tank. The extenders are attached to the bombs on the left inboard wing pylon.

67-287, without its sharkmouth, is carrying 12 MK82 low drag 500 pound bombs. The sharkmouths painted on the Korat aircraft were against regulation, so they were painted over whenever the 388th wing was expecting an inspection.

66-8738 (FP 407th TFS) leads the Korat flight in a descent to the release altitude for a LORAN pathfinder release. 66-8738 was shot down over North Vietnam on October 5, 1972.

In this view the wing man is flying a formation landing approach. Two ship formation landings were often practiced which ensure the aircrews could fly them safely when needed in an emergency.

SAM Hunter-Killer Missions

The SAM hunter-killer mission was a modification of the SAM IRON HAND strike mission. Two F-4Es teamed up with two F-105G Wild Weasels. The F-4E aircraft were armed with CBUs. The F-105s fired their missiles at the SA-2 FANGSONG radars. The F-4s used the explosion of the Weasel's missile to locate the SAM site. The F-4s then destroyed the missile launchers and missiles with CBUs.

The flexibility of the CBU equipped F-4Es taking part in hunter-killer missions was displayed on a LINEBACKER mission of September 29, 1972. Part of Condor flight's mission description stated: "Condor flight reported two SAMs launched from VN-159 (site identification number) directed toward Condor 01. Condor 01 and 02 fired four AGM-45s on two FANSONG radars. Condor 03 and 04 (F-4Es) expended their CBUs on Phu Yen Air Field and reported visual BDA (bomb damage assessment) of one MiG-19 and two MiG-21s on fire and four MiG-21s damaged."

67-274 accelerates for takeoff carrying CBUs and MK82s. The stabilator is positioned with its leading edge down, in position to lift the nose when the aircraft reaches takeoff speed.

67-290 pulling into a second row revetment following a combat mission. The air refueling receptacle can be seen extended on the back of the aircraft.

F-105G 63-8321 and F-4E 67-288 half of a SAM hunter-killer team patrol over North Vietnam.

The weapons load on this F-4E is made up of two CBUs on each inboard wing pylon and six MK82 500 pound bombs on the centerline.

67-287 seen here with its North Vietnam strike load of 12 MK82 500 pound bombs, two AIM-7 Sparrows aft, and an ECM pod in the left forward Sparrow well

Interdiction Missions

Interdiction missions were flown in an attempt to disrupt the flow of materiel into North Vietnam from Communist China, and into South Vietnam from North Vietnam. These missions targeted both the transportation system (railroads, roads, bridges, etc.) and the materiel being transported. The ordnance was varied and based on the target type. If the target was the system itself, 500 and 750 pound general purpose bombs were normally used. If the target was the materiel being transported, 500 and 750 pound general purpose bombs were fitted with 3 foot long fuze extenders, and used along with cluster bombs (CBUs) to destroy the materiel. The fuze extenders caused the bombs to detonate just before they impacted the ground, causing maximum damage to supplies which might be stacked on the ground in the target areas.

The interdiction missions flown in North Vietnam were generally against specific targets such as railroad yards, bridges or river ferry points. On 11 June 1972, as part of LINEBACKER operations, I took part in a flight of four F-4Es which was targeted against traffic on the northeast railroad (the railroad which runs from Hanoi, northeast to the Chinese border). The targeted section of railroad ran along the floor of a valley between two steep parallel ridge lines. Anti-aircraft guns were set up along the tops of these ridges. In order to attack the railroad effectively, the F-4s flew parallel to the tops of the ridges, exposing the aircraft to a cross fire from the AAA on both ridges. Over the target area, the four ship formation entered a wheel pattern, at an altitude which was out of range of the AAA in the target area. In the wheel pattern the aircraft flew in a circular path with the spacing between aircraft varying. The aircraft then took turns diving out of this wheel pattern at different positions in the wheel, as a result, the aircraft attcked the target from different headings each turn. The different attack headings made targeting of the aircraft by the ground gunners more difficult. Three bombs were released on each pass using the dive toss mode. After release, the F-4 pulled back up into the wheel. This mission was quite successful, resulting in numerous rail line cuts, leaving damaged and destroyed rail cars scattered throughout the target area, and causing numerous secondary explosions.

During early LINEBACKER operations Korat F-4Es flew interdiction missions against ferry points on the Dong Hao river which runs very close to the DMZ (demilitarized zone) separating North and South Vietnam. Earlier strikes had destroyed the bridges forcing the North Vietnamese to use barges as ferries to move trucks and supplies across the river. The trucks backed up along roads awaiting their turn to cross the river by ferry were a very lucrative target.

The missions flown in Laos against the Ho Chi Minh trail normally had no specific targets, but instead were directed by a Forward Air Controller (FAC) flying an OV-10 (NAIL FAC) in the lower threat areas, or an F-4 (fast FAC such as the 388th's Tiger FACs) in higher threat areas. The flight, after inflight refueling, would "check in" by radio with the ABCCC C-130E orbiting on station and receive a target area and the call sign of the FAC working in that area. The F-4 flights then proceeded to the area and contacted the FAC. The FAC located specific targets in the area and then marked the location using white phosphorus smoke rockets directing the flights where to bomb.

This close up of 67-385 weapons load shows the 12 MK82 low drag 500 pound bombs and the two AIM-7E missiles.

This KC-135A, assigned to the 4104 Air Refueling Squadron (Provisional) at Korat, is being towed back to its parking space.

Above and below: OV-10 Broncos like these performed as forward air controllers (FACs) for most of the interdiction and strike missions in Laos. (Don Logan Collection)

Fast FAC (Tiger FAC)

F-4Es from Korat flew as forward air controllers (FACs), mostly in southern Laos. The 388th's fast FAC's were called Tiger FACs. Selected crews from both the 34th and the 469th TFS were qualified as Tiger FACs. It was their job to fly low and fast over enemy territory where operation of slower flying FAC aircraft such as OV-10s and O-2s could not be done safely. FACs located targets such as enemy troop concentrations, supplies, or trucks and other equipment. After locating the targets, the fast FACs marked the targets in the same manner as other FACs using white phosphorus smoke rockets. The FACs directed the strike aircraft where to release their weapons based on the position of the smoke from the rocket fired by the FAC.

This F-4D 66-7659, from the 433rd TFS, 8th TFW at Ubon RTAFB, loaded with rocket pods on the inboard wing stations and an M61 20mm gun pod on the centerline, was serving as a fast FAC for a Korat F-4E fourship when this photo was taken in April of 1972.

Close Air Support

Korat F-4s flew close air support missions in support of friendly troops in contact (TIC) with enemy forces. Because of Korat's geographic location in Thailand, not South Vietnam, close air support missions were normally flown in northern Laos supporting friendly Laotian forces, leaving missions in South Vietnam to the U.S. forces located there. One exception to this was the Viet Cong attack on the city of An Loc in South Vietnam. 34th and 469th TFS F-4Es flew missions in support of the battle for An Loc. A four ship flight departed Korat, air refueled with KC-135s along the eastern boarder of Thailand, and then attacked enemy targets in the An Loc area. FACs on the ground directed the flights to specific targets using smoke in a manner similar to that used by the airborne FACs.

After expending the ordnance in the vicinity of An Loc, the F-4s landed at Bien Hoa Air Base, approximately 100 miles to the southeast. At Bien Hoa the aircraft were rearmed and refueled. The same crews then returned to the An Loc area in support of the battle. Ordnance for the missions from Bien Hoa to An Loc was "Snake and Nape" (MK-82SE Snakeye high drag 500 pound bombs and BLU-27 Napalm fire bombs) with three MK-82SEs on both of the inboard wing

A VNAF A-1H comes in for a landing at Bien Hoa Air Base South Vietnam as a four ship outbound for a strike at An Loc awaits its turn for takeoff.

pylons and two BLU-27s on the centerline MER. In this case the snake and nape was used as an anti-personnel weapon to prevent the Viet Cong from overrunning friendly positions. After releasing the ordnance the F-4s rendezvoused with KC-135s to air refuel, and then returned to Korat.

Wonder Arch revetments like this one were used at Bien Hoa Air Base to protect the aircraft from Viet Cong rockets.

This Korat F-4 is parked at Bien Hoa. The aircraft is loaded with MK82 Snakeyes on the inboard wing stations and BLU-27 Napalm bombs on the centerline station in support of the battle at An Loc.

This Korat F-4 parked in a "wonder arch" revetment at Bien Hoa prior to launch in support of the battle at An Loc.

67-276 here on the boom was assigned to the 34th TFS along with the MiG Killer 69-276.

Over northern Laos, 67-287 heads for the target area.

The F-4E in this photo (67-317), taken in March of 1972, was carrying markings of the 421st TFS (LC tail code), 366th TFW based at Da Nang Air Base South Vietnam. The aircraft, with the prototype ASX-1 TISEO (target identification system, electro-optical) installed, was in SEA performing operational tests on the system.

69-7551 refuels on the boom while 69-7267 awaits its turn.

Search and Rescue (SAR)

When an aircraft was shot down or crashed, search and rescue (SAR) of the aircrew became the highest priority. Airborne strike aircraft were diverted to assist in the SAR effort. My first combat mission, originally scheduled as a trail interdiction mission into the border area between southern Laos and South Vietnam, turned into a SAR mission. After delivering our ordnance against a truck park and storage area, out flight of two F-4Es was contacted by the on station ABCCC C-130 and informed of a SAR effort in our vicinity. A USAF pilot flying an A-37, acting as an advisor to friendly forces, had been shot down and had ejected inside enemy territory. My pilot informed the ABCCC that our flight had expended our bombs, but had not expended any of our 20mm ammunition in the nose gun. Since we were the closest to the location of the downed pilot we were given the coordinates and proceeded to the location.

Upon arrival in the area we were able to locate the downed pilot in a small stand of trees in the center of some rice paddies. The enemy forces were in a treeline about 200 yards from the downed pilot. Our flight of two positioned itself so that one of the aircraft was always flying over the downed pilot towards the enemy. If the enemy forces tried to cross the rice paddy in the direction of the downed pilot we strafed the area they were trying to cross. We continued to cover the downed pilot in this manner until friendly forces in an armored personnel carrier were able to recover the pilot. The pilot was then taken to a local friendly village where he was picked up by helicopter.

Chaff Dispensing

F-4 aircraft were used to lay chaff corridors as part of the support of air strikes into high threat areas. This chaff, loaded into MJU-1 chaff bombs (approximately the same size and shape of a M117 750 pound bomb), was designed to block enemy radar reception and shield the strike flights from SA-2 SAMs. Each of the F-4s carried 12 chaff bombs. For adequate protection, the strike aircraft had to be within the chaff clouds in both azimuth and elevation. This protected the aircraft from detection by enemy radar sites on the ground. Chaff clouds are not solid targets, but they did not allow radar energy to pass through them to be reflected by the aircraft back to the radar. If the aircraft were outside the chaff, they could be detected and displayed on the enemy radar scope allowing them to be targeted for attack by enemy defenses.

In order to dispense the protective chaff corridor for the strike force, the flight preceded the strike flights into the tar-

67-316, seen here on April 7, 1972 flying as lead of a chaff dispensing flight is carrying MJU-1B chaff bombs, visible on the centerline rack, behind the wing tank.

get area. The formation used to dispense the chaff required flying four aircraft at medium altitude, in a line abreast formation, with 4,000 feet wing tip to wing tip spacing between the aircraft. While flying straight and level each aircraft released one chaff bomb in unison every 20 seconds. This resulted in a chaff corridor about three to four miles wide, and about 35 miles long.

On 7 April 1972, taking part in the first attack in North Vietnam as part of the renewed bombing operations of North Vietnam called FREEDOM TRAIN, I was flying as the WSO in the lead aircraft given the mission of dispensing a protective chaff corridor for the strike flights. Laying the chaff presented a problem as our flight was highlighted as we released the chaff, flying at the leading edge of this growing corridor. As a result, an estimated 40 SA-2 SAMs were fired at our flight. There were no hits; however, at one point two missiles detonated simultaneously at our altitude, both about 400 feet away, one at my 3 o'clock position, and the other at about 8 o'clock position.

During LINEBACKER, F-4s based at Udorn RTAFB equipped with ALE-38 chaff dispensers, were used for chaff support. Eight chaff bombers were required to lay down the chaff for the B-52 strikes. Eight F-4s using the ALE-38 could lay down a continuous chaff corridor five miles wide and 105 miles long, sufficient to cover the entire ingress and egress routes within North Vietnam. Two flights of four sowed two chaff corridors for the B-52s. These corridors provided the approach and departure corridor. The chaff was dispensed at the B-52's route altitude of 36,000 feet. In order to screen the B-52 force and prevent their detection, it was necessary for the chaff flights to know the exact planned tracks of the B-52 cells.

Many link ups with aircraft from other wings occurred on the tanker. Two 8th TFW F-4Ds, 65-609 (FA) from the 25th TFS on the boom and 65-793 (FG) from the 433rd TFS complete their pre-strike refueling. After refueling, 793 equipped with LORAN lead the Korat flight on a pathfinder weapon delivery.

One of Korat's two newest F-4Es was 69-7267.

ESCORT

On escort missions the F-4's flew with the aircraft we were escorting, AC-130 gunships, B-52s, or other tactical strike aircraft. It was the escorts job to protect the aircraft they were escorting from an enemy threat. The primary threat on AC-130 Spectre escort missions was from ground fire, while the primary threat on B-52 or strike flights was enemy MiGs.

AC-130 Escort

The AC-130 Escort missions were night missions, taking off after dark and returning before sunrise. The normal mission duration for escorting an AC-130 (Spectre gunship) was four hours, one hour to fly from Korat to the area in which the gunship was working, two hours with the gunship, and after being relieved by another flight, one hour back to Korat. The aircraft configuration for Spectre escort missions was a 270 gallon wing tank on each outboard wing station, three CBUs on each inboard wing pylon, and two M36E2 Thermite bombs on the centerline. The longest duration combat mission I flew was as Spectre escort. This mission lasted almost eight hours. On this specific mission, the relief flight aborted and did not show up, so we had to remain with the gunship until he had finished his mission.

To escort an AC-130, which spent most of its time at about 8,000 feet above the ground in a 45 degree bank (to keep its sensors and guns pointed at the target area on the ground), the F-4 at about 12,000 feet had to be in a 45 to 60 degree bank to keep position on the gunship. Two F-4s were used for each escort mission. One remained on station with the gunship while the other cycled back to a KC-135 for refueling. When the wingman returned from the tanker, he relieved the F-4 escorting the gunship, who then cycled to the tanker, and in turn returned to relieve the wingman. This continued

This view shows the normal strike load carried by Korat's F-4Es. The load consisted of 12 MK82 500 pound bombs and two AIM-7 Sparrows. The empty ECM Pod rack can be seen in the left forward Sparrow well.

until the flight was relieved by the next flight, or the gunship finished its mission and left for its home base.

If the F-4 detected gunfire directed at the gunship, the F-4 crew informed the gunship and attacked the area from which the gunfire was detected. The size and brightness of muzzle flash of the gun gave an indication of what size gun was firing. If it appeared to be small arms or machine guns, then CBUs were used to suppress the fire. If it appeared to be a large caliber gun (37 mm or bigger), the Thermite bomb was used. Before dropping the Thermite bomb, the F-4E crew gave a radio call to the gunship telling the gunship that they were "in hot with a funny bomb." This told the gunship crew to turn off the light sensitive night vision sensors which could be damaged by the very bright light caused by the Thermite explosion and fire.

B-52 Escort

During the late winter and early spring of 1972, the North Vietnamese moved some of their MiG-21 fighters to airfields in southern North Vietnam. As a result, B-52s on ARC LIGHT missions in northern South Vietnam were vulnerable to attack from these MiGs. As a result, F-4s were assigned the duty of escorting the Buffs and if the B-52s were attacked by MiGs, the F-4s were to engage the MiGs and prevent them from reaching the bombers. The F-4 aircraft configuration for B-52 and Strike Escorts was "three bags" (both 270 gallon wing tanks and the 600 gallon centerline tank), two AIM-9 Sidewinders on each inboard wing pylon, two AIM-7E2 Sparrows in the aft Sparrow wells, ECM pods in the forward Sparrow wells, and of course a loaded nose gun. If, while in the B-52/Strike escort configuration, an engagement with enemy fighters was anticipated all three external fuel tanks could be jettisoned. I flew three B-52 escort missions, all with no threats to the Buffs ever materializing.

During LINEBACKER II F-4s from Korat and Udorn flew as escort for the B-52s, while the MiGCAP missions were flown by F-4s from Udorn. The escort formations flew behind, below, and slightly offset from the B-52s. This was the position the MiG had to reach for a successful Atoll missile launch (the Atoll was Soviet copy of the U.S. AIM-9B Sidewinder heat seeking missile). When attacking an aircraft from the rear, the missile's effective range was approximately one and one half miles. With the B-52s flying at an altitude between 35,000 and 40,000 feet. The F-4s flew five miles in radar trail from the last B-52 in each cell, offset to the side at an altitude of about 25,000. In order to reach the Atoll firing envelope, the MiG had to perform a high-speed snap-up maneuver from the rear. Accomplishing this maneuver, the MiGs, highly visible in full afterburner, had to position themselves in front of the F-4s. The F-4s could then launch their AIM-9s at the MiGs. Radar guided AIM-7s were not be used in order to ensure against the possibility of the AIM-7s accidently transferring to home-on-jam mode and homing in on the B-52 jamming systems. The escorts were assigned to protect multiple B-52 cells. The escorts flew an elongated racetrack orbit pattern parallel to the B-52 flight path. They followed the first B-52 cell to the target, then after making a 180 degree turn outbound, picked up the next following B-52 cell and escorted it to the target.

Above and below: The bombs visible here on 67-379 were specially painted for this mission on Christmas Day 1971.

Strike Escort

A four ship of F-4s accompanied the strike aircraft into the target area to serve as a strike escort. A North Vietnamese tactic was to engage the strike flights with MiGs prior to the strike flight reaching the target. In order for the flight to engage the MiGs, for self defense, they would have to jettison their bombs. The MiGs then retreated leaving the strike flight with no bombs to attack the target. For strike escort missions, F-4s loaded with air to air ordnance accompanied strike loaded F-4s carrying air to ground ordnance. The theory behind strike escort was; if the strike flight was "jumped" (attacked) by MiGs, the strike aircraft retained their bombs and proceeded to the target while the strike escort aircraft, loaded for air to air, would engage the MiGs.

COUNTER AIR

Counter Air, or CAP (Combat Air Patrol) missions were intended to seek out enemy aircraft and engage and destroy them before they became a threat to other aircraft. The aircraft configuration for CAP missions was the same as for B-52 or Strike escort missions.

Barrier (BAR) CAP

When performing a BAR CAP the flight of aircraft positioned itself between the aircraft they were protecting (Tanker Force,

These two F-4Ds from 432nd Tactical Reconnaissance Wing at Udorn, 66-8707 from the 13th TFS and 66-8737 from the 555th TFS, are carrying an air to air load with four AIM-7s each.

DISCO AEW&CS aircraft etc.) and the enemy, and aggressively attacked and pursued any attacking enemy aircraft long before they could get close enough to pose a threat.

MiG CAP

A MiG CAP or MiG Sweep mission was designed to clear airspace of enemy aircraft. The CAP aircraft flew to an assigned area including areas such as enemy airfields and patroled the area until enemy aircraft were detected. The CAP then engaged and attempted to destroy those enemy aircraft.

Below: When flying as ICEMAN 03 on June 21, 1972, 67-283 escorting a chaff flight, became the first 388th TFW aircraft to shoot down a MiG in 1972.

69-276 seen here after my return from a Strike Escort mission. We jettisoned all three tanks attempting, unsuccessfully, to catch a MiG-21. The four AIM-9s, two on each inboard wing pylon were carried as part of the escort weapons load.

The early morning sun lights 67-379 as it takes on fuel for a LINEBACKER strike.

This 555 TFS F-4D (66-8737) at Udorn RTAFB is equipped with ARN 82 LORAN as evidenced by the "towel rack" antenna on the F-4s back.

This LORAN equipped 13th TFS F-4D (66-8785) is serving as pathfinder for a LORAN delivery over Northern Laos.

This 13th TFS F-4D (66-7531), releasing MK82s, some with fuze extenders, is part of a Udorn F-4D four ship which combined with four Korat F-4Es for a eight ship LORAN release over Northern Laos.

This Triple Nickel (555 TFS) F-4D (66-8747), seen prior to entering North Vietnam on a LINEBACKER mission, is loaded for MiGs carrying four AIM-7 Sparrows visible in the front and rear missile wells, and four AIM-9s on the inboard wing pylons, blocked from view by outboard wing tanks on the inboard wing pylons.

An F-4E of the 4th TFS, 366th TFW at Da Nang Air Base, South Vietnam closes from pre-contact position while a Korat F-4E (67-269) waits its turn. The 366th TFW was the other wing based in SEA operating F-4Es.

67-269 with flaps and gear down on approach for landing at Korat.

67-275 in its revetment at Korat.

This JJ tail coded 34th TFS F-4E (67-269) is carrying 12 MK82 low drag bombs. The bombs on the inboard pylons have fuze extenders installed.

Though Korat's F-4Es normally carried their AIM-7 Sparrow missiles in the two rear missile wells, the F-4E in this photo is carrying one AIM-7 in the left forward missile well.

The lengthened nose gun shroud seen in this photo was added to all of Korat's F-4Es during January and February 1972. The wedge shaped strike camera used for photographing battle damage, can be seen below the right engine intake, installed in the right forward missile well.

The fuze extenders on the wing mounted MK82 are visible on this 34 TFS F-4E.

This photo of 67-379 was taken in May of 1972 after the new gun fairing was added to the 20mm nose gun.

This distinctive sharkmouth identified Korat's F-4Es. This sharkmouth was copied by many other F-4E units.

68-313 heading east on an interdiction mission to southern Laos.

67-379 carrying 12 MK82 bombs. The three bombs on each inboard wing pylon are fitted with fuze extenders.

67-287 waits its turn as an F-4D from Ubon moves away from the tanker.

F-105G Wild Weasel Missions

The Wild Weasels' primary mission was Surface To Air Missile (SAM) suppression. These SAM suppression missions used the code name IRON HAND. To accomplish SAM suppression, the 388th TFW had F-105G aircraft assigned as Weasels. The primary targets of the Wild Weasels were the SA-2 SAM sites. The sites were located using the emissions of the site's FANSONG radar system. The FANSONG radar was used by the enemy to locate the airborne target and direct the SA-2 missile in flight. The Weasel backseater, known as a "Bear", directed the pilot to position his aircraft in order to fire an Anti-Radiation Missile (ARM) at the source of the radar emissions. The AGM-45 Shrike and AGM-78B Standard ARM were the ARMs used by the F-105G Weasels. An ARM homed in on the radar emissions and, if successful, destroyed the radar antenna and the radar van on which the antenna was mounted. The destruction of the radar put the site "out of commission" until another radar system could be set up.

Strike Flight Escort Missions

During strike escort missions the IRON HAND F-105Gs accompanied the strike force into the target area and fired their ARMs at any fire control radar which was detected. As a result of accurate firing of the ARMs, the North Vietnamese modified their tactics. The North Vietnamese SAM crews fired barrages of up to six unguided SAMs with out turning on their fire control radars. Without the fire control radars broadcasting, the ARMs did not have a target to home on. Using this tactic, even though the SAMs were not guided, the enemy hoped the Strike force would jettison their ordnance prior to the target. In any case this tactic distracted and harassed the aircrews during the final stages of the strike.

SAM Strike Missions

A SAM strike mission consisted of a four-ship of F-105Gs whose mission was to fly over SAM defended areas and "troll" for SAMs. If a fire control radar "came up" the Weasels fired at the site's radar.

SAM Hunter Killer Missions

The SAM hunter-killer mission was a modification of the SAM IRON HAND strike mission and was the most effective Wild Weasel mission type flown during LINEBACKER and LINEBACKER II. In the hunter-killer missions two F-105G Wild Weasels teamed up with two F-4E strike aircraft armed with CBUs. The intent was to find and destroy SA-2 SAM sites. The Wild Weasel was the hunter and used its equipment to locate the active SAM sites. The Weasel then fired a missile at the SAM site radar, the F-4s used the explosion of the Weasel's missile to locate the SAM site. The F-4s then "killed" the missile launchers and missiles with CBUs.

LINEBACKER II B-52 SAM Suppression Missions

Although the Wild Weasel SAM hunter-killer teams worked extremely well during daylight and in clear weather conditions, SAM suppression support for the B-52s during LINEBACKER II had to rely on different tactics. The weather over the Hanoi/Haiphong area of North Vietnam during LINEBACKER II operations was expected to include a think undercast ranging from 3000 to 8000 feet. This thick undercast in the threat zones forced the mission planners of the 388th TFW to revert to the IRON HAND missions. The procedure developed for B-52 SAM suppression used IRON HAND aircraft (F-105Gs and F-4Cs) in flights of four escorting the B-

Silhouetted against a thunderstorm this F-105G Wild Weasel prepares to move into precontact position for air refueling.

52s in very close proximity. This required a great deal of pre-mission coordination so that the Weasels could position themselves between the flight path of the B-52s, and the known ground threats. The Weasels set up in a race track orbit alongside the B-52's flight path. The legs of the race track were planned so that two Weasels were always headed into the target area while two were headed outbound. This ensured two Weasels were always pointing toward the highest threats and were ready to attack any SAM sites turning on their radars to acquire the attacking forces.

63-8345 is parked in one of the taxi through revetments on the Korat flightline.

62-4423 on takeoff at Korat.

These 561st TFS F-105Gs (MD tail code), recently arrived from McConnell AFB, are being prepared for combat on the Korat flight line just after their arrival on April 6, 1972.

This 561st TFS F-105G (63-8342) was shot down over North Vietnam on April 15, 1972. Both crewmembers were not recovered and were declared MIA (Missing In Action).

Above and below: 62-8320 on takeoff at Korat. During 1967 8320 claimed three MIGs while assigned to the 388th TFW at Korat. The claimed kills could not be confirmed.

62-4423 at liftoff.

63-8316 here with a JB tail code. 17th WWS aircraft carried ZB tail codes into mid 1972 when they were changed to JB.

63-8265 an F-105G from the 561st TFS, 23 TFW at McConnell AFB, Kansas. The 561st TSF deployed to Korat in April of 1972 as part of the build up of forces called Constant Guard.

In this low view of a 561st TFS F-105G, the ECM fairings on the side of the fuselage are visible along with an AGM-45 on the left wing and an AGM-78 on the right wing.

EB-66 Missions

The EB-66 aircraft were used as a radar systems and communications jamming platform with the assigned mission of blocking the North Vietnamese communications and early warning network, thereby neutralizing their defensive system. Active ECM activity was conducted by the EB-66 aircraft.

During LINEBACKER and LINEBACKER II missions, specific orbits were established for each mission to optimize ECM coverage during ingress and egress of the strike force. Three EB-66s were normally on station during a mission. Two provided active ECM functions, and the third was an airborne spare orbiting over Laos or the Gulf of Tonkin providing contingency support. The EB-66 ECM support was sometimes supplemented with Marine and Navy EA-6s.

During LINEBACKER operations the North Vietnamese moved some SAM sites beneath the western orbits of the EB-66s, forcing the EB-66s to move further away from the target areas. As a result of the increased distance, the effectiveness of the EB-66s declined. Often EB-66 ECM support was not used during LINEBACKER missions, because of the vulnerability of the EB-66s and the size of the effort required to protect them from MiGs. On many strike missions the ECM protection was provided by the strike aircraft themselves. During LINEBACKER II, after a two month bombing lull in the Hanoi/Haiphong area of North Vietnam, intelligence data on SAM site locations was poor and inaccurate. As a result, the confirmed operating areas became suspected operating areas, and as the operating rules allowed EB-66 operation over suspected SAM operating areas, three EB-66 aircraft flew in an orbit forty miles west of Hanoi to provide ECM support to the B-52s. Two were primary aircraft, and the third was assigned as a spare.

EC-121 "DISCO" AEW&CS Mission

These AEW&CS aircraft (call sign DISCO) monitored North Vietnamese aircraft movements and communications, and directed the U.S. MIGCAP flight intercepts. During LINEBACKER missions, EC-121s were on station, initially over Northern Laos, and later over the Gulf of Tonkin to ensure coverage of all North Vietnamese airspace, flying over 13,931 combat sorties during the Vietnam war. They issued 3,297 MiG warnings, assisted in 25 MiG shootdowns, and also assisted in 80 SAR missions for recovery of downed aircrew.

Due to the limited U.S. radar warning and control capability in and around the areas north and west of Hanoi, the U.S. forces flying over North Vietnam during the first few months of LINEBACKER were highly vulnerable to surprise MiG attacks. A new early warning system, code named TEABALL was established at Nakhon Phanom RTAFB. This system was designed to provide positive control and near real-time MiG warnings to the strike forces. DISCO aircraft had the responsibility for controlling intercepts, issuing MiG alerts, warning pilots of potential border violations with China, and sometimes issuing SAM warnings.

Initially the DISCO aircraft maintained orbits over Laos providing MiG tracking coverage as far into North Vietnam as possible, with the Navy providing MiG traffic and warning information from a Navy cruiser (call sign RED CROWN) on station in the Gulf of Tonkin (its station is shown on the air refueling track map as the box next to item 24). RED CROWN tracked the MiGs and provided information on their location to the MiGCAP flights. This ability was limited when tracking MiGs operating at low altitude, or when the number of both friendly and enemy aircraft was so large that it saturated RED CROWN's tracking capabilities. A DISCO orbit over the Gulf of Tonkin was established in July, after two F-4s (call sign BASS 02 and BASS 04) were shot down by low flying MiGs eluding detection by RED CROWN. I was flying in the second of the two F-4s shot down. The new DISCO orbit over the Gulf was able to supplement the RED CROWN coverage and offered some degree of improvement.

Positive results from the TEABALL control system using information from the EC-121s in orbit and RED CROWN was

EB-66E 54-523 in partial 42nd TEWS markings sits in the arming area awaiting departure. (USAF)

evident almost immediately after July. Between February and July, 18 U.S. tactical aircraft were lost to MiG attacks while U.S. forces destroyed 24 MiGs. After TEABALL became operational, U.S. losses between August and October numbered only five, in comparison with 19 MiGs shot down.

54-435 taxis for takeoff.

AGE (Auxiliary Ground Equipment) used for engine start, painted yellow, sits off each wing of 54-523.

The antennas on the bottom of this EB-66E (54-479) were dedicated jamming enemy electronic emissions.

EC-121R Missions

The EC-121Rs were used as radio relay aircraft for the sensors "seeded" on the Ho Chi Minh trail. These sensors detected any movement or activity in their vicinity. The Bat Cats flew along the trail receiving the transmissions from the seeded sensors on the ground. The transmissions were amplified and relayed to monitoring stations in Thailand. The transmissions were then analyzed and used to determine enemy troop and materiel movements, and set requirements for air strikes.

The BAT CAT EC-121s were used to relay the signals received from the sensors placed along the Ho Chi Minh trail by U.S. forces. (USAF)

C-130E ABCCC Mission

C-130E ABCCC (Airborne Battlefield Command and Control Center) aircraft were first assigned to Korat on April 30, 1972. The ABCCC mission was to manage tactical air resources, and coordinate target requirements with airborne strike flights and ground forces. The ABCCC aircraft flew using the call signs MOONBEAM, ALLEYCAT, CRICKET, and HILLSBORO.

HC-130P "KING" Search and Rescue Mission

These aircraft were modified HC-130H with equipment added to refuel helicopters in flight. They flew using the call sign

KING and were used as the on scene commander for all SAR missions in Southeast Asia directing the SANDY SAR aircraft and the rescue helicopters. If additional tactical air support was required, the KING aircraft coordinated their needs with the ABCCC aircraft. The ABCCC then assigned tactical air resources to the SAR effort with the KING aircraft taking over control of the resources once they were assigned. In addition, the KING birds served as the inflight refueling tankers for the HH-53 Super Jolly Green Giant rescue helicopters used by the SAR force.

Below: This EC-121R, in its revetment is being prepared for its next flight.

Above and below: HC-130P takes off from Korat.

HH-53 Jolly Green Giant rescue helicopters, like this one on Korat's ramp were the helicopter used in rescue efforts during 1972. The helicopter was armed with a 7.62 mini-gun in each forward window and one on the aft ramp.

KC-135A Tanker Operations

Because of Korat's geographic position in the middle of Thailand, far away from most target areas, inflight refueling from Young Tiger KC-135As was a necessity if the fighters were to reach their targets.

In flight refuelings took place in planned refueling areas or refueling tracks as they were called. The accompanying map shows the refueling tracks located in Southeast Asia. The refueling tracks each had names; the tracks which ran in a north-south direction had colors for names (Black, Green, Orange, Red, White, Blue, Purple, Tan, and Yellow), the tracks which ran east west had fruits for names (Lemon, Peach, and Cherry). Black, Green, Orange, Lemon, White, Purple and Tan refueling tracks had an anchor (race track orbit pattern) at each end of a long refueling track, while the rest of the tracks consisted of a single anchor with no long track.

The key point of an anchor refueling area was called the Air Refueling Control Point (ARCP) as shown in the refueling diagrams. For a Point Rendezvous, the tanker would normally arrive at the ARCP at least 15 minutes prior to the expected receivers, and upon reaching the point would start a left hand turn to fly an orbit pattern that resembled a left hand race track (this left hand orbit pattern was called an anchor). The tanker commenced its final elongated lap just as the receivers entered the air refueling track at the opposite end at an altitude slightly lower than the tanker, in an ideal position to refuel. When the receivers closed to a slant range of 21 miles with the tanker 26 degrees to the left, the tanker would begin a 180 turn in order to role out about one half mile in front of the receivers. If possible, the tankers and receivers tried to complete the refueling while flying straight on the elongated portion of the anchor.

When large numbers of tactical aircraft were involved in coordinated mass strike forces, some aircraft loitered with the tanker force until the strike fore was assembled. In these cases the tankers would "Top Off" the strike aircraft, so that the entire strike force headed into the target area with full tanks. Most refueling areas had an additional less elongated race track with an ARCP downstream (towards the target area). This additional race track was called an anchor orbit (items 13, 14, 15, & 16 on the air refueling track map). The same anchors were generally used for pre-strike and post-strike refuelings, with the post-strike refueling portion on the side of the anchor opposite that of the pre-strike refueling.

If a quick rendezvous was necessary due to receiver running low on fuel, a technique called "head-on-rendezvous" (shown in the air refueling diagrams) was used. In this rendezvous the receiver and the tanker headed directly at each other. At the proper point, the tanker turned 90 degrees to the right, and then a 90 degree turn to the left, offsetting the tanker from the receiver. When the receivers closed to a slant range of 21 miles with the tanker 26 degrees to the left, the tanker would begin a 180 degree turn to roll out directly in front of the receiver.

When single flights of aircraft required refueling they would rendezvous with a tanker already "in orbit" in the anchor. The flight received radar vectors (headings and altitudes to fly) from ground radar sites in Thailand and South Vietnam, and close on the tanker until they visually acquired the tanker. The F-4s modified this procedure, and used their onboard air intercept radar to accomplish a "fighter turnon rendezvous." In a "fighter turnon", the F-4 flew directly at the tanker, picked up the tanker on the intercept radar and made a sweeping 180 degree turn to roll out 1/2 mile behind the tanker.

Tanker refuelings were of two types; boom refueling, and probe and drogue. The KC-135 is fitted with a "flying boom" through which fuel is pumped under pressure to the receiver aircraft. The boom is fitted to the aft of the aircraft. It is rigid and telescopes to allow it to be lengthened and shortened, and has two wing like control surfaces called ruddervators mounted in a "V" at the trailing end of the rigid boom. The boom can be configured, while on the ground for probe and drogue retueling. To accomplish probe and drogue refuelings, a flexible hose and a conical shaped metal cage (called a drogue) is added to the nozzle at the end of the boom. While configured for probe and drogue, the boom cannot be used for boom and receiver refuelings.

The receiver aircraft have a receiver receptacle, usually on the top of the fuselage near the cockpit, which is designed to connect the boom to the receiver's fuel system. Using the ruddervators the boom operator, laying in the aft of the tanker, flies the boom to the receiver aircraft receptacle, and by extending the boom latches the boom nozzle into the receiver receptacle. Once latched fuel flows at up to 900 gallons per minute into the receiving aircraft. When the receiver's tanks had filled, or when the boom disconnected from the receiver, the fuel flow automatically stopped. Korat's F-4s, A-7s, and F-105Gs refueled using the flying boom.

For probe and drogue refuelings, the boom is lowered and extended, allowing the drogue to trail behind it. The receiver aircraft has a probe attached to the aircraft somewhere forward of the cockpit. The receiver pilot then flies the aircraft so that the probe catches the drogue, and when latched, fuel is then pumped through the boom to the probe, and into the receivers fuel system. Korat's CD-00s used the probe and drogue system for refuelings. The F-105Gs were also capable of probe and drogue refueling using a retractable probe located above the gun in the aircraft nose. The F-105s usually refueled using their receptacle and the tanker's flying boom.

AIR REFUELING TRACKS IN SOUTHEAST ASIA

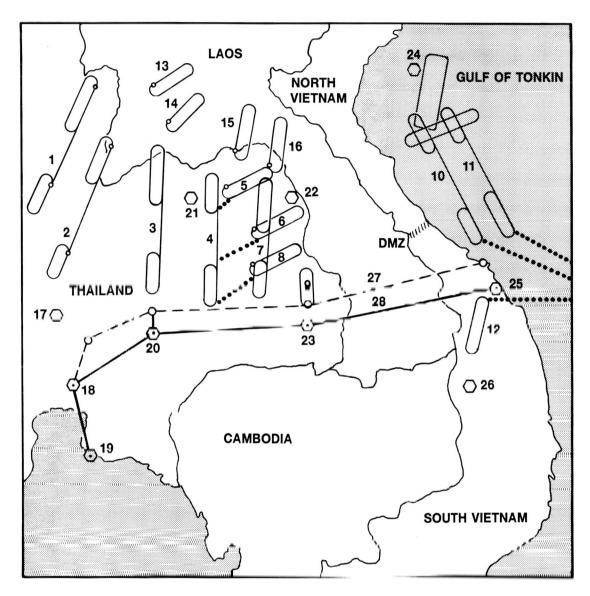

REFUELING TRACKS
1   BLACK AIR REFUELING TRACK
2   GREEN AIR REFUELING TRACK
3   ORANGE AIR REFUELING TRACK
4   RED AIR REFUELING TRACK
5   LEMON AIR REFUELING TRACK
6   PEACH AIR REFUELING TRACK
7   WHITE AIR REFUELING TRACK
8   CHERRY AIR REFUELING TRACK
9   BLUE AIR REFUELING TRACK
10 PURPLE AIR REFUELING TRACK
11 TAN AIR REFUELING TRACK
12 YELLOW AIR REFUELING TRACK

ANCHOR REFUELING POINT EXTENSIONS
13 GREEN AIR REFUELING ANCHOR ORBIT
14 ORANGE AIR REFUELING ANCHOR ORBIT

15 RED AIR REFUELING ANCHOR ORBIT
16 WHITE AIR REFUELING ANCHOR ORBIT

AIR BASES/COMMUNICATION SITES USED FOR TANKER RENDEZVOUS
17 TAKHLI/ALPHA LIMA
18 DON MUANG (BANGKOK)
19 U-TAPAO
20 KORAT/DRESSY LADY
21 UDORN BRIGHAM
22 NAKHON PHANOM/INVERT
23 UBON/LION
24 RED CROWN NAVY CRUISER
25 DA NANG/PANAMA
26 PLEIKU/PEACOCK

FLIGHT ROUTES ACROSS SOUTHEAST ASIA
27 AMBER 8
28 KILO CHARLIE

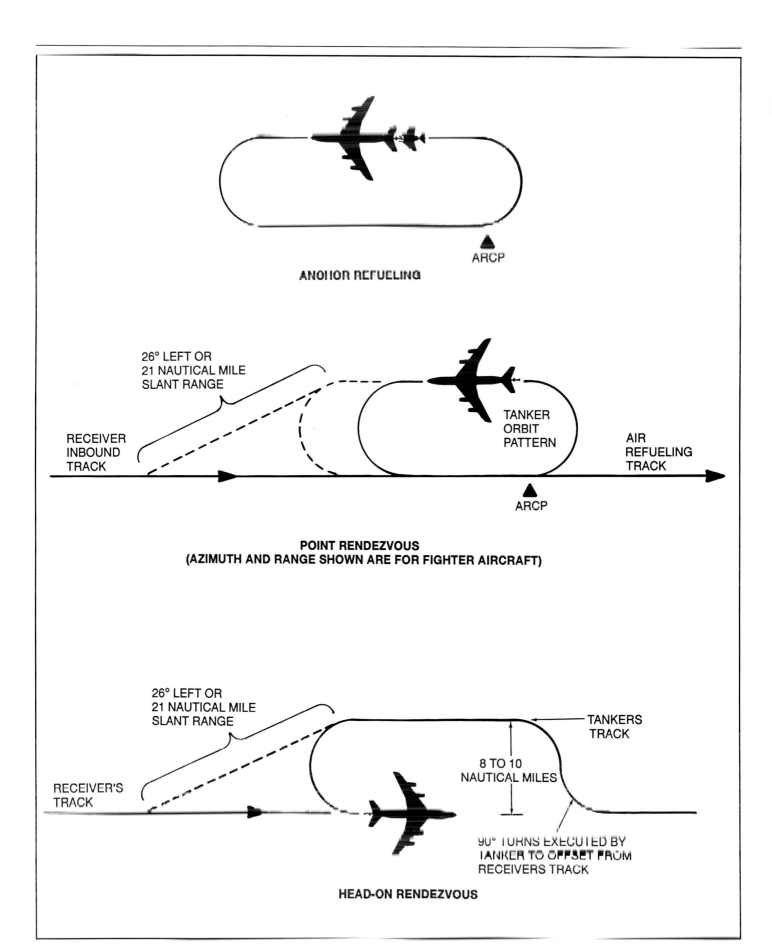

ANCHOR REFUELING

26° LEFT OR
21 NAUTICAL MILE
SLANT RANGE

RECEIVER
INBOUND
TRACK

TANKER
ORBIT
PATTERN

AIR
REFUELING
TRACK

ARCP

POINT RENDEZVOUS
(AZIMUTH AND RANGE SHOWN ARE FOR FIGHTER AIRCRAFT)

26° LEFT OR
21 NAUTICAL MILE
SLANT RANGE

TANKERS
TRACK

8 TO 10
NAUTICAL MILES

RECEIVER'S
TRACK

90° TURNS EXECUTED BY
TANKER TO OFFSET FROM
RECEIVERS TRACK

HEAD-ON RENDEZVOUS

The need for inflight refueling was common to all F-4s. This tanker already had a 49th TFW F-4D and a 14th TRS RF-4C "in tow" when the four ship of 469th TFS F-4Es arrived.

EB-66E 54-523 is refueling through the refueling probe which is plugged in to basket being trailed by KC-135A 57-2598. (USAF)

62-4416 receives fuel from KC-135A 63-7989. (USAF)

This is the view of a KC-135A tanker seen by the receiver aircrew as they move from the tankers wing to the precontact position.

This receivers view of the boom shows the nozzle on the end of the boom. This nozzle plugs into the receiver aircraft's refueling receptacle

This receptacle on the back of an F-4E, receives the nozzle and latches it in place before fuel is transferred.

F-4E 66-322 for its prestrike refueling is taking on fuel from KC-135A 56-3608.

Loaded with 12 MK82s, 34th TFS F-4E 67-208 receives fuel from KC-135A 57-1435.

Because of Korat's position in the center of Thailand, without KC-135A tankers like 60-0349 seen here, the F-4s, F-105s, and A-7s from Korat did not have the range to strike targets in North Vietnam.

388TH TACTICAL FIGHTER WING MIG KILLS - 1972

The fighting arm of the Air Force of the Democratic Republic of Vietnam (North Vietnamese Air Force) was made up of Soviet manufactured MiG fighters. When LINEBACKER started the North Vietnamese Air Force's 204 MiGs were made up of the following aircraft:

AIR FIELD	MiG-21	MiG-19	MiG-15/-17	TOTAL
Bai Thoung	2			2
Dong Suong			12	12
Kep	46	1		47
Kien An	12			12
Phuc Yen	8		49	57
Quan Lang			3	3
Yen Bai	25	32	14	71
TOTALS	93	33	78	204

On December 18, 1972, at the beginning of LINEBACKER II, the MiGs had been reduced to 145. They were made up of the following:

AIR FIELD	MiG-21	MiG-19	MiG-15/-17	TOTAL
Dong Soung	13			13
Gia Lam	4	2	5	11
Hoa Lac	2	3		5
Kep	4	3	27	34
Kien An			4	4
Phuc Yen	8	4	8	20
Quan Lang	1			1
Yen Bai	7	28	22	57
TOTALS	39	40	66	145

June 21, 1972

The first 388th MiG kill of 1972 took place on the 21st of June, when a flight of four F-4Es (ICEMAN flight) from the 469th TFS at Korat escorted two flights of chaff dispensing aircraft over Route Package 6 (Hanoi area) in North Vietnam. Two MiG-21s engaged the U.S. aircraft, one attacking the chaff force and the other pursuing the lead Phantom, flown by Col. Mele Vojvodich, Jr., and Major Robert Maltbie.

"I saw three different MiGs and got off a shot at one of them. I didn't see the missile impact because I was distracted by a MIG-21 on my right," Vojvodich commented.

F-4E 67-283, the aircraft flying in #3 position, crewed by Lt.Col. Von Christiansen, the 469th TFS operations officer, and Major Kaye Harden, presently never volunteered from being shot down.

"We were flying escort for two flights of chaff dispensing aircraft on 21 June 1972," reported Christiansen, "when at least two MiG-21 aircraft attacked the chaff force. At about 0649Z, two MiG-21 aircraft were initially sighted at 12 o'clock high to the chaff force, crossing our egress course from left to right. At this time the MiG-21s were two to three thousand feet above the chaff force, partially obscured by a 500-foot-thick broken overcast cloud layer. The chaff force was positioned less than 100 feet below the base of the overcast. As the MiGs came abreast of the chaff force, they executed a hard nose low turn to the left, quickly positioning at 6 o'clock on the chaff force and the lead MiG-21 commenced an attack. While following his leader through the turn, the number two MiG appeared to sight Vojvodich and his wingman below him. He then pulled high momentarily to gain a favorable position and initiated an attack on the two F-4s. Possibly because our element was positioned high on the left in fluid-four formation, it appeared that the number two MiG did not see me and my wingman. Upon observing him rapidly closing at 6 o'clock on Vojvodich and his wingman, we called them to break left. The MiG's rate of closure was such that he continued nearly straight ahead after firing two Atoll missiles at ICEMAN 2. ICEMAN 2 managed to evade both missiles with a hard turn to the left. By going to maximum power and performing an acceleration maneuver, we were able to stabilize our position at 5,000-6,000 feet behind the number two MiG in a slight descending turn. He was in afterburner power."

"After acquiring a full system radar lock-on, we attempted to fire two AIM-7 missiles, but neither AIM-7 missile launched. We then switched to heat and picked up a strong IR (infrared) tone from our second AIM-9 missile when the number two MiG was positioned in the gunsight reticle. Three AIM-9 missiles were ripple-fired at the MiG, who was in a level, gentle bank to the left. The first missile appeared to guide normally, but detonated about 50 feet right of the MiG's tail. Major Harden observed the second missile guide directly into the MiGs tail, causing the aircraft to explode and burn fiercely from the canopy aft. The pilot ejected immediately and was observed to have a yellow parachute. I did not observe the second AIM-9 impact on the MiG, because I immediately transferred my attention to the number one MiG, which was pulling off high after attacking an F-4 of another flight."

"We initiated a maximum power pull-up toward the number one MiG and thereafter maneuvered with him at very high speed until achieving a position at his 6 o'clock. During this time, the number one MiG executed numerous evasive maneuvers while descending from 20,000 feet to 1,000 feet as we closed for a gun attack. Radar lock-on was obtained and although tracking was by no means perfect, firing was initi-

92 • THE 388th TACTICAL FIGHTER WING

NORTH VIETNAMESE MIG BASES

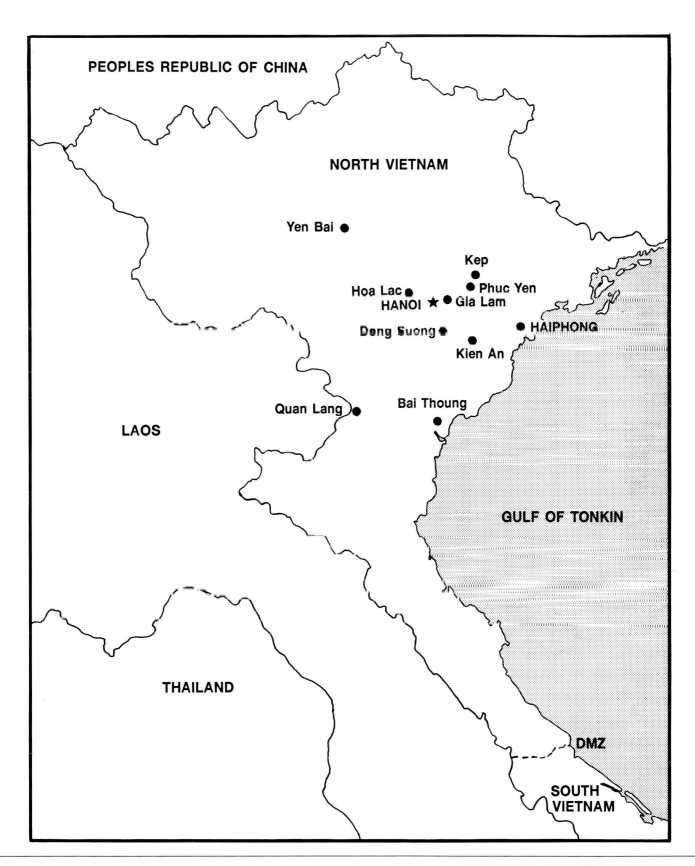

PEOPLES REPUBLIC OF CHINA

NORTH VIETNAM

Yen Bai ●

Kep
●
Hoa Lac ●   Phuc Yen
●
HANOI ★ ● Gia Lam

Dong Suong ●
●
Kien An

● HAIPHONG

LAOS

Quan Lang ● Bai Thoung
●

GULF OF TONKIN

THAILAND

DMZ

SOUTH
VIETNAM

67-283 was the first Korat aircraft to score a MiG kill during LINEBACKER when it downed a MiG-21 with a sidewinder (AIM-9) on June 21, 1972.

ated from about 3,000 feet with a short burst. Thereafter, we fired several short bursts while slowly closing range and attempting to refine the tracking solution. Suspecting a gunsight lead prediction problem, we began to aim slightly in front of the MiG and observed strikes on the left wing just as the gun fired out. The engagement was terminated due to Bingo fuel state at that time."

"Colonel Christiansen saved the men in our lead aircraft by telling them to break just at the right time,"Major Harden reported. "Two missiles from a MiG exploded close behind them. We turned into the low MiG and fired two Sidewinders. One of them knocked the tail section off of the MiG and the pilot ejected. The aircraft spun to the ground in flames."

In addition, there was another MiG damaged in the engagement. This marked the first confirmed victory by a 388th TFW aircraft since 23 August 1967.

September 3, 1972

The next 388th MiG kill occurred on September 3. The kill was made by an F-4E (67-392) of the 388th TFW. The F-4E was part of EAGLE flight, a SAM Hunter-killer flight made up of two F-4Es (EAGLE 03 and 04) and two F-105Gs (EAGLE 01 and 02). They were flying SAM suppression in the vicinity of Phuc Yen airfield. After the flight had evaded 20 SA-2s fired at them during the 12 and 1/2 minutes they had spent in the target area, EAGLE 02 sighted an Atoll missile heading

This star was hastily painted on 67-283 after its MiG kill on June 21, 1972.

towards EAGLE 01. EAGLE 01's aircraft commander, Major Thomas Coady, flying with Major Harold Kurz, put the aircraft into a hard right turn, causing the missile to miss the Thunderchief's left wing by approximately 20 feet. As he broke right he visually acquired a light blue MiG-19. The MiG pilot then pressed a cannon attack against EAGLE 02, also an F-105 and crewed by Major Edward Cleveland and Captain Michael O'Brien. A hard right turn also saved them. The MiG had a very high rate of speed and overshot. As the MiG barrel-rolled away from Eagle 01, it passed over aircraft 3, an F-4E flown by Major Jon Lucas and 1LT. Douglas Malloy. Flying inverted, the MiG headed east, probably trying for Phuc Yen Air Field.

"He came in from our 1 o'clock position," said Major Lucas, "and I started a left turn to maneuver into firing position. The MiG then started a left-descending turn at which time I called for an auto-acquisition. The weapon systems officer, Lieutenant Malloy, went to boresight and confirmed the switch settings. I hit the auto-acquisition switch with the MiG-19 framed in the reticle. Lieutenant Malloy confirmed a good lock-on. I counted 4 seconds and squeezed the trigger. The left aft missile light went out, indicating expenditure of an AIM-7. I started to select Master Arm and Guns to follow up with a gun attack. At that time, approximately 0440Z, a SAM was observed tracking our aircraft, and a turn was initiated into the SAM to negate its track. We then turned back towards the MiG and observed a pastel orange parachute with a man hanging in the harness. Missile impact was not observed due to the turn into the SAM, but Cleveland and his wingman called

the MiG-19 burning and spiraling towards the ground and also observed the parachute.

September 12, 1972

On September 12 aircrews of the 388th TFW downed three MiG-21s. The aircraft credited with MiG kills were F-4Es 67-279 and 67-268, and F-4D 65-608. Two MiGs were destroyed by aircrews in a flight of four F-4Es (FINCH flight) escorting chaff flights northeast of Hanoi, in the vicinity of Kep airfield. Three or four MiGs came in from 4 to 6 o'clock and attacked one of the chaff flights as it approached the target area. The lead F-4, crewed by Lt.Col. Lyle Beckers and 1LT. Thomas Griffin, observed a MiG aligning itself to the rear of the chaff flight from which point he could launch a missile. According to Becker's account, "I obtained an auto-acquisition lock-on and attempted to fire two AIM-7 missiles. The MiG-21 fired an Atoll missile at the chaff flight and broke straight down. I pursued and fired two AIM-9 missiles, one of which impacted the MiGs left wing. Flames and smoke were observed coming from the left wing. I then selected guns and proceeded to fire 520 rounds of HEI tracer. Projectile impacts and additional fire were observed on the fuselage of the MiG."

The MiG-21 was last observed in a steep descent, burning. An Atoll missile from the MiG, however, found its mark and destroyed one of the chaff aircraft before Beckers and his WSO could drive him off. This aircraft was an F-4E (69-7266) assigned to the 335th TFS, 8 TFW flying out of Ubon

67-283 sits in its revetment with the MiG kill star painted on the intake splitter.

RTAFB. The aircrew, Captain Fredrick McMurray the pilot, and Captain Rudolph Zuberbuhler the WSO, successfully ejected, but were captured and held as POWs until released following the signing of the Peace Treaty with North Vietnam

Meanwhile, Major Gary Retterbush with 1LT. Daniel Autrey in the back seat of FINCH 03 attacked another MiG-21.

"We turned into the MiGs and accomplished a radar lock-on," Retterbush reported. "Two AIM-7s were fired but did not guide. Three AIM-9s were fired, but missed by a matter of feet. We then closed on and downed a MiG-21 with 20mm cannon, firing approximately 350 rounds. The 20mm with tracer was observed impacting the fuselage, wing, and canopy, causing fire and smoke."

The MiG went into an uncontrolled climb with its nose 65 degrees up, slowed to 150 knots, then dropped. Major Retterbush reported that as the MiG dropped past him he saw the pilot slumped forward in the cockpit. The cannon had found its mark. As the F-4Es left the battle area, they observed a smoke trail and a large fireball.

Later in the day, another F-4 flight (ROBIN flight) from the 388th TFW (aircraft 1, 3, and 4 were F-4Es, aircraft 2 was an F-4D), while escorting a strike flight attacking the Tuan Quan railroad bridge, encountered two attacking MiG-21s. The first MiG appeared in an 8 o'clock position and lined up on the strike flight. The lead aircraft fired one AIM-7 missile ballistically to distract the MiG, then turned in pursuit as the MiG broke away. This "shot across the bow" detonated about 1000 feet in front of the Phantom. In hot pursuit, the flight

67-368 (FINCH 03) was one of three 388th F-4s to down a MiG kills on September 12, 1972.

leader then fired another AIM-7, followed by three AIM-9 missiles. They all missed. The second AIM-7 detonated 500 feet from the target, and the nearest AIM-9 detonated about 200 feet from the MiG.

The ROBIN 02, piloted by Captain Michael Mahaffey, with 1LT. George Shields in the rear seat, had better luck with its ordnance during the engagement. As the flight leader was chasing the first MiG, a second MiG-21 dropped between the two F-4s. "It went right across in front of us." Mahaffey later commented, "and it looked a lot bigger than I thought a MiG was supposed to look. We rolled right, tracked, and fired one AIM-9 which guided and impacted the MiG in the tail section, blowing off parts of the aircraft. The MiG went into a spin

67-392 on October 6 1972, along with 66-313, was credited with downing its second MiG, a MiG-19, by causing the MiG to fly into the ground.

67-268 seen here illuminated by an early morning sun was credited with shooting down a MiG-21 on September 12, 1972.

On October 15, 67-301, seen here in March of 1972 was credited with shooting down a MiG-21 using an AIM-9 Sidewinder. This was the 388th TFW's last MiG kill of the Vietnam war.

from 16,000 feet and more pieces fell off the aircraft. It was last seen in a spin below 8,000 feet, about 20 nautical miles southwest of Yen Bai Air Field.

October 5, 1972

The first aerial victory for the month of October, came on 5 October when MiGs from Kep airfield opposed a strike force. An Air Force escort flight of F-4Es (ROBIN flight) from the 388th TFW engaged the enemy in a heated battle. A MiG-21 was downed by Captain Richard Coe and 1LT. Omil Webb III in the lead aircraft (ROBIN 01) F-4E 68-493. Coe reports:

"We received vectors from DISCO (EC-121) for two MiGs off Bullseye on ingress to initial point. They seemed to be heading in our direction. DISCO gave continuous vectors until the flight we were escorting called MiGs at 0 o'clock high. The formation began a hard left turn. After two turns I observed two MiG-21s in route formation at 10 o'clock high. at about three miles and heading 280 degrees. We began a lazy one-G descending turn to get to 6 o'clock. The auto-acquisition switch was activated with the MiGs still in the pipper. I then fired one AIM-7. At this time someone called, "Someone has a MiG at 6 o'clock, tracking." We rolled up to check 6 o'clock. I then checked 12 o'clock where I saw a smoke trail entering a black smoke cloud and a large white column exiting the other side. We then broke hard right and on roll-out observed the white column leading down to two large dirt clouds rising from the ground.

October 6, 1972

The following day, October 6, two F-4E aircrews of EAGLE flight, a SAM hunter-killer team, destroyed a MiG-19. The aircraft were 67-0392 and 66-0313.

The first Phantom was manned by Major Gordon Clouser and 1LT. Cecil Brunson, and the other by Captain Charles Barton and 1LT. George Watson. The manner in which this MiG was destroyed was unusual. DISCO warned the flight of approaching MiGs. The flight was then in the vicinity of Thai Nguyen. The F-105 flight leader and his wingman, making up the other half of the flight, moved out of the area as prebriefed while Clouser and Barton turned to make contact with the enemy. Clouser then observed a MiG-21 sliding into a 7 o'clock position, Barton observed a MiG-19 attempting to achieve a 6 o'clock position on the element. Clouser called a hard left break to provide self-protection for the Phantoms and to divert the MiGs from the F-105s. Because of the ordnance on board, the maneuverability of the F-4Es was limited, and therefore they jettisoned the ordnance and fuel tanks. The MiGs were dangerously close to a firing position and the two backseaters, Brunson and Watson, warned their pilots of the danger. To disrupt enemy tracking, Barton went into a vertical dive in afterburner with a weaving pattern. Meanwhile, Clouser was able to maneuver out of the MiGs range without resorting to a dive. The MiG-19 pilot followed Barton's aircraft, its guns blazing, and Clouser rolled in behind the MiG to create a sandwich. The MiG-21 sandwiched Clouser, creating an F-4/MiG-19/F-4/MiG-21 chain. Barton continued the

66-313, an F-4E of the 469th TFS, taxis back to its parking spot after a mission. On October 6, 1972, along with F-4E 67-392 making up the F-4 half of a SAM hunter-killer four ship, 66-313 was credited with destroying a MiG-19.

69-276, flying here with 67-208, scored a gun kill of a MiG-21 on October 8, 1972, after the F-4s AIM-9s failed to launch.

dive and bottomed out at 300 feet above a valley floor between two mountain peaks.

The MiG-19 pilot was apparently so engrossed with the chase that he failed to notice the vertical dive angle until it was too late. His aircraft impacted with the ground. Both F-4Es recovered and the MiG-21 hastily withdrew from the battle. Each F-4E crewmember was subsequently credited with one-half of a MiG kill.

October 8, 1972

The next 388th claimed MiG fell on 8 October. Major Gary Retterbush and Captain Robert Jasperson crewed the lead aircraft of LARK flight (69-276), a flight of F-4Es dispatched by the 388th TFW. This was Major Retterbush's second MiG, the first occurring on September 12. The following is the Major's account:

"On 8 October 1972, while flying lead on a strike escort mission, we received warnings of MiGs coming in from the north. We jettisoned our tanks and maneuvered behind a MiG-21 who began evasive action. Our infrared missiles (AIM-9 Sidewinders) failed to fire, so we closed and fired the 20mm cannon. Several good hits were observed and the MiG burst into flames. The pilot ejected at approximately 1,500 feet before the aircraft impacted the ground."

October 15, 1972

The last 388th TFW MiG kill of the Vietnam war occurred on October 15. The "last" kill was credited to Major Robert Holtz and his WSO, 1LT. William Diehl flying F-4E 67-301 as PAR-ROT 03. The flight had been dispatched by the 388th Wing to escort three flights of F-4 strike aircraft to the vicinity of Viet Tri. Numerous MiGs were engaged by this flight before Holtz finally downed one of them. His claim statement provides a record of their activities: "I and my GIB (Guy In Back) engaged a number of MiG-21s in the vicinity of Viet Tri and succeeded in destroying one MiG-21. While escorting a strike package of three flights of F-4s from Ubon we were vectored by RED CROWN to two MiGs in my 12 o'clock position. These bandits were picked up visually at about two miles and a hard left turn was made to engage as they passed overhead and away at a rapid rate. Seeing that these two were no longer a threat, we started to return to escort duties when my wingman saw and engaged another MiG-21 with myself flying fighting wing. This MIG headed for the clouds and disappeared. At this time the strike flights were too far ahead of us to catch, so I called for an orbit in the vicinity of Viet Tri to cover the strike flights on egress. While in this orbit my wingman and I re-engaged one more time each, with negative results. We got separated by numerous F-4s going through our flight after another MiG. While in a right hand turn to rejoin my wingman, I circled a cloud and noted a white parachute about

66-313, seen here loaded with CBUs on the inboards and MK82s on the centerline, on October 6, 1972, along with 67-392 scored a MiG kill by causing a MiG-19 to fly into the ground, killing the pilot and destroying the aircraft.

3.000 to 4,000 feet AGL. At this time I told Lieutenant Diehl to mark the time and position in case it was one of our pilots. I then noted a silver MiG-21 orbiting the descending parachute about the same altitude (3,000 to 4,000 feet) and within a 30 degree cone of my nose to the right. The MiG was not maneuvering but instead was in a lazy right bank of about 20 degree and about 3,000 feet ahead. I fired an AIM-9 which

came off the rail, did a slow roll and then went straight up the MiGs tail and exploded, blowing pieces of tail section and almost one complete elevator off the aircraft. The MiG rolled violently to the right and started towards the ground, nose down at about 20 degree to 30 degree and on fire. At this time I disengaged and egressed the area."

DATE	MIG Type	U.S. A/C	Serial # Number	Weapon Used	TFS (Callsign)	CREW Members
21 June	MiG-21	F-4E	07-200	AIM 0	469th (ICEMAN 03)	LC V CHRISTIANSEN MAJ KAYE HARDEN
03 Sept	MiG-19	F-4E	67-392	AIM-7	34th 35th (EAGLE 03)	MAJ JON LUCAS 1LT DOUGLAS MALLOY
12 Sept	MiG-21	F-4E	67-275	AIM-9/20mm	35th (FINCH 01)	LC LYLE BECKERS 1LT THOMAS GRIFFIN
12 Sept	MiG-21	F-4E	67-268	20mm	35th (FINCH 03)	MAJ GARY RETTERBUSH 1LT DANIEL AUTREY
12 Sept	MiG-21	F-4D	65-608	AIM-9	469th (ROBIN 02)	CPT MICHAEL MAHAFFEY 1LT GEORGE SHIELDS
5 Oct	MiG-21	F-4E	68-493	AIM-7	34th (ROBIN 01)	CPT RICHARD COE 1LT OMRI WEBB III
6 Oct	MiG-19	F-4E	67-392	MANEUVERING	34th (EAGLE 03)	MAJ GORDON CLAUSEN 1LT CECIL BRUNSON
		F-4E	66-313	MANEUVERING	34th (EAGLE 04)	CPT CHARLES BARTON 1LT GEORGE WATSON
8 Oct	MiG-21	F-4E	69-276	20mm	35th (LARK 01)	MAJ GARY RUTTERBUSH CPT ROBERT JASPERSON
15 Oct	MiG-21	F-4E	67-301	AIM-9	34th (PARROT 03)	MAJ ROBERT HOLTZ 1LT WILLIAM DIEHL

KORAT AIRCRAFT LOSSES - 1972

During 1972, 19 Korat-based aircraft were destroyed, 16 due to combat and three in landing accidents. One A-7D, two EB-66s, eight F-105s, one F-4D, and seven F-4Es were lost.

A-7D Losses

One A-7 was lost while units of the 354th TFW were stationed at Korat. 71-310 (callsign SLAM 04) was lost on December 24, as the result of a mid-air collision with Raven 21 (an O-2 FAC) over the target (a North Vietnamese AAA gun position) located in Laos. The pilot, Captain C. Reiss ejected, was captured, and later rescued. The O-2 pilot was not recovered and was listed as MIA (Missing In Action).

EB-66 Losses

Bat 21 - April 2, 1972

In spite of the resumption of offensive operations against North Vietnam on April 1, 1972, the only EB-66 combat loss in 1972 actually took place over northern South Vietnam when on April 2, 1972 an EB-66C (54-466), call sign "Bat 21", was brought down by a SAM fired across the DMZ. Only one of the crew, Lt.Col. Iceal E. (Gene) Hambleton was rescued, the other crewmembers were declared MIA.

On the afternoon of April 2, 1972, two EB-66s (Bat 21 and Bat 22) were escorting a cell of B-52s bombing near the DMZ. Bat 21, an EB-66C, was carrying its normal six man crew consisting of a pilot, navigator, and four Electronic Warfare Officers (EWOs). Three SAMs were fired at Bat 21 from inside South Vietnam. One of the missiles scored a direct hit on Bat 21, impacting the aircraft in the EWOs compartment area. The aircraft went down with only the navigator Lt.Col. Hambleton being able to eject.

As Hambleton descended in his parachute, he was able to talk to Captain Jimmie Kempton, a FAC flying an OV-10 based at DaNang AB, South Vietnam. The area where Hambleton was going to land was completely blanketed by clouds. Capt. Kempton descended under the clouds and located Hambleton visually as he floated down in his chute.

Coincidentally a airborne SAR force was in the area after their mission to evacuate some U.S. Army advisors from Quang Tri had been canceled. Two of the A-1s, SANDY 07 and SANDY 08, heard the emergency radio calls and headed for Hambletons position. The SANDYs spent the next few hours taking to Hambleton on the radio and trying to keep the North Vietnamese from capturing him. Hambleton was able to direct the SANDYs where to release their ordnance in order to keep the enemy away from him.

Meanwhile, Kempton had flown southward, trying to contact someone who could go in and pick up Hambleton. He was able to contact four U.S. Army helicopters, two UH-1B

A four ship of 354 TFW on the ramp at Korat being prepared for a strike mission. A-7Ds like these replaces the A-1s on "SANDY" search and rescue missions during the second half of 1972. (Don Logan Collection)

A-1Hs and A-1Js like this one belonging to the 1st Special Operations Squadron (SOS), 56 Special Operations Wing at Nakhon Phanom RTAFB, carrying a TC tail code flew as "Sandy" SAR support aircraft until being sent to the South Vietnamese Air Force during 1972. (Dan Logan Collection)

Cobra gunships and two UH-1H passenger carrying "slicks." Approaching Hambletons position, two of the four helicopters were shot down. One UH-1H was completely destroyed with no survivors. The Cobra, Blue Ghost 28, was able to reach the beach where its crewmembers were rescued. With darkness fast approaching, there was no pickup for Hambleton that night.

At 2100 hours, Nail 59, from Nakhon Phanom (NKP), relieved the DaNang FAC. Nail 59 was a new version of the OV-10 aircraft with the Pave Nail system installed. The Pave Nail system used Loran equipment to provide accurate survivor position at night or in bad weather. Nail 59 made contact with Hambleton and was able to establish his exact position using radio bearings from Hambleton. Hambleton was about 1000 meters north of the town of Cam Lo on the north bank of the Mien Giang River.

Throughout the night, FACs from DaNang and NKP maintained continuous patrol over Hambleton. Shortly before dawn on April 3, two Pave Nail OV-10s from NKP were enroute to arrive over Bat 21 at first light. Hambleton had taken refuge in a big clump of bushes surrounded by a large field, completely surrounded by enemy troops.

Because of the foul weather, visual airstrikes were not possible. Using Loran coordinates supplied by the two Pave Nail OV-10s, Task Force Alpha targeting center at NKP put together the information necessary for Loran strikes in Hambleton's area. The data was then passed to the Nail FACs on the scene. The FACs directed F-4s, using the F-4s Loran equipment, to place their ordnance around Hambleton's position.

On the second day of the operation, an SA-2 (SAM) claimed Nail 28. The missile, coming out of the clouds directly below Nail 28. The OV-10 had no time to evade the missile. It impacted the OV-10 in the tail booms, blowing the aircraft apart. The crew members, Bill Henderson and Mark Clark, both ejected from the aircraft. Clark landed south of the Cam Lo river, while Henderson landed in a field about 500 meters from Bat 21. He was able to hide in a clump of bamboo until he was captured when the North Vietnamese cut down the bamboo to use as camouflage. Captain Henderson was taken north to Hanoi and held as a POW until his release at the end of the war. Clark found a barbed wire enclosed area and figured it was a good place to hide. For the rest of that day and the next two days, FACs and fighters worked the area over.

On April 6, the first attempt was made to pick up the two survivors. Jolly 62, the rescue helicopter, was escorted by a flight of A-1E SANDYs. The plan was to run in over Clark's position on a northwest heading, cross the river and the road, pick up Hambleton, make a left turn, pick up Clark, and exit the area.

Jolly 62 safely crossed the river, but as they started to head for Bat 21, they came under fire from a village. The Jolly turned right and headed right for the village. This right turn put them into even more heavy machine gun fire. The helicopter was taking numerous hits. The SANDY pilots watched as Jolly 62 limped back across the river, but, as described by one of the SANDY pilots, "a flame shot out below the main rotor, the helicopter nosed up, rolled left 90 degrees, and pieces starting falling from the aircraft. It hit the ground on its left side, the fire continued to burn finally consuming the entire aircraft."

For the next several days the air forces kept the pressure on the whole area. The two survivors continued to hide and evade while remaining in contact using their survival radios. It was determined that it was too risky to try another pick up by air. In order for the two survivors to be picked up on the ground they were to move down the river and meet up with a Marine ground team.

After receiving word of the new plan, Clark and Hambleton started working their way down river to meet the Marines. Clark being closer to the river swam and floated down the river and met the first Marine team. Hambleton, having to move through a mile of mine field, took over four hours to reach the river. After resting, he crossed the river, found a log, and began floating down the river. After dawn he hid in the foliage along the river bank. For three nights Hambleton made his way down the river, his progress constantly followed by the FACs. On the fourth day he sighted a sampan; using his pre-arranged signal he called out to the sampan. The Marines on board acknowledged his signal and took him aboard. Bat 21 returned to Korat after his 12 day ordeal.

EB-66 Crashes At Korat - December 24, 1972

The last EB-66 loss, an EB-66E, occurred on December 24, Christmas Eve. It was the result of a crash on landing accident at Korat when both engines failed. All three crewmembers were killed. The two mission controllers (EWOs) were killed in the crash. The pilot ejected when the aircraft was in a near inverted attitude and was killed. The total number of EB-66s

lost during the Southeast Asia War was fifteen, six in combat and nine in operational accidents.

Below: This EB-66C (54-466), flying with the callsign BAT 21, was shot down on April 2, 1972 near the DMZ (border between North and South Vietnam) starting a rescue effort which lasted 12 days.

F-105G Losses

Eight F-105G Wild Weasels were lost during 1972 of which seven were combat losses, and one the result of a landing accident. SAM missiles caused five of the combat losses, one was shot down by an Atoll from a MiG-21, and one went down as a result of a malfunction of the Weasel's own missile. Of the sixteen crewmembers involved, two were listed as KIA (killed in action), three MIA, five POW, and six rescued.

February 17 - Triple 3 Is Shotdown

63-8333, flying as part of JUNIOR flight was hit over North Vietnam, probably by a SAM, and crashed at sea. The two crewmembers, Captain Jim Cutter the pilot, and Captain Ken Fraser the backseater, were captured and held as POWs. They were returned to the U.S. following the signing of the peace treaty.

April 15 - A SAM Claims Another Weasel

63-8342 was hit by a SAM while attacking a SAM support site over North Vietnam. It was not known whether the crew was able to eject. Both crew members, Pilot Captain A. Maleja and backseater Captain O. Jones were listed as MIA.

May 11 - A MiG Bags ICEBAG

While flying as part ICEBAG flight on SAM suppression mission, 62-4424 was hit by an Atoll missile from a MiG-21 as it attacked a FANSONG radar. The crew, pilot Major Bill Talley and backseater Major Jim Padgett, ejected. They were both captured and spent the remainder of the war as POWs.

May 17 - Crashed On Landing

63-8347 was destroyed by fire after it ran off the runway on landing as the result of a blown main landing gear tire. The aircraft landed heavyweight after having returned to Korat

On February 17, 1972 63-8333 was hit over North Vietnam and was able to make it out to sea before the crew had to eject. The crew was captured and held as POWs.

63-8342, belonging to the 561st TFS, was shot down on April 15, 1972 nine days after deploying to Korat.

KORAT AIRCRAFT LOSS LOCATIONS

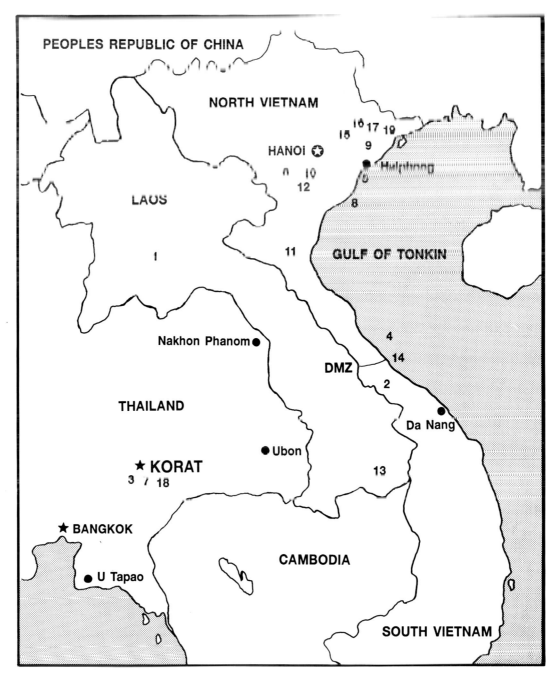

1	A-7D	71-310	24 DEC 72	11	F-105G	63-8359	16 NOV 72
2	EB-66C	54-466	2 APR 72	12	F-4D	66-265	20 JUL 72
3	EB-66E	UNKNOWN	24 DEC 72	13	F-4E	66-371	10 FEB 72
4	F-105G	63-8333	17 FEB 72	14	F-4E	67-303	8 JUN 72
5	F-105G	63-8342	15 APR 72	15	F-4E	67-277	1 JUL 72
6	F-105G	62-4424	11 MAY 72	16	F-4E	67-296	5 JUL 72
7	F-105G	UNKNOWN	17 MAY 72	17	F-4E	67-339	5 JUL 72
8	F-105G	62-4443	29 JUL 72	18	F-4E	67-385	22 SEP 72
9	F-105G	63-8360	17 SEP 72	19	F-4E	69-276	12 OCT 72
10	F-105G	63-8302	29 SEP 72				

following an aborted mission. Both crewmembers were recovered.

July 29 - A Coordinated MiG Attack And A Bad AGM Claims An F-105

During a day SAM suppression mission 63-4443 was lost. F-4E 66-0367 of the 4th TFS, 366 TFW had just been shot down by a MiG-21, 4443 was performing a defensive MiG break (Split S) when its own missile damaged its right wing. The aircraft was able to make "feet wet" where the crew ejected. Pilot Major T. Coady and backseater Major H. Murphy were rescued by U.S. Marine helicopters.

September 17 - The Crew Of Two Is Killed

While performing an armed reconnaissance mission as part of Condor flight, 62-8360 was shot down. The flight was attacking a FANSONG radar when 8360 was hit by a SAM. The aircraft crashed at sea. Both crewmembers, Captain T Zorn and 1LT M. Turose, were killed.

September 29 - One Captured, One MIA

63-8302 was shot down by a SAM while attacking a FANSONG radar site as part of CROW flight on a SAM suppression mission. Both crewmembers ejected, Lt.Col. Jim O'Neil was captured and held as a POW until the end of the war. Captain Mike Bosiljevac fate was unknown and he was listed as MIA.

November 16 - Escorting B-52s Can Be Dangerous

On November 16 Bobbin 05 (F-105G 63-8359) was flying a night escort for a B-52 strike near Thanh Hoa when it was hit by a SAM. The pilot Major Norman Maier and his Bear, Captain Kenneth Thaete were able to eject from the burning aircraft, setting up a dramatic rescue accomplished with help from 354th TFW A-7s stationed at Korat. One of the B-52 crewmembers, Major Peter Giroux (later shot down and held as a POW), witnessed the shootdown and said:

"It was a remarkably clear night at altitude with our cell attacking targets on a west to east heading. We had a F-105 Wild Weasel as escort, among others. He was flying off our left wing with another aircraft also in our area. Beneath them was a thick undercast. Suddenly a bright circle lit up on the ground as a SAM left its launcher. It slipped out of the overcast and moved toward the lead aircraft (B-52) in our cell, but then missed as it passed in front of him. A second or two later I saw a second SAM light up the overcast almost directly below the F-105. It popped through the clouds and almost immediately struck the underside of the Thud. The ejection seats went out almost immediately and I was surprised that I could see them fire at this distance. We continued the bomb run still straight and level and made our release."

The two crewmembers drifted down separately, and did not see each other land. Captain Thaele landed in the flatlands and headed for a nearby ridgeline to hide. Major Maier landed near a road, where he immediately abandoned his gear and headed for cover up a hill. The local North Vietnamese troops knew the flyers were in the area and tried to flush them out by occasionally firing their guns. The high grass where the airmen were hiding made it very difficult for the enemy troops to find them.

63-8302 seen here in March of 1972, was shot down by a SAM on September 29, 1972 while flying a SAM suppression mission.

With bad weather closing in, a rescue plan had to be developed quickly. A SAR force of about 75 aircraft, including A-7 SANDYs, EB-66s, F-4s, F-105Gs, HC-130s, EC-121s, and HH-53 Super Jolly Green Giant helicopters was put together for an initial rescue attempt. This attempt was to take place a dawn. The on-the-scene commander flying an A-7D (70 0070) with the call sign SANDY Lead was Major Colin Clark of the 356th TFS, 354th TFW stationed at Korat. Major Clark had never before flown in a SAR, but had experience in rescues, He had been shot down and rescued twice before while flying F-100 early in the conflict. In fact, in late August 1964, he became the first U.S. Air Force pilot to be shot down in the Vietnam War.

The command post for this SAR mission was a 56th Air Rescue and Recovery Squadron HC-130 "Kingbird," also from Korat. At daybreak the SANDYs rendezvoused with the Jolly Greens (from Nakhon Phanom) above a solid overcast along the Laos-North Vietnam border. Major Clark looked for a hole in the overcast through which the HH-53s could let down. He located the two downed airmen by using the A-7s on board computer to pinpoint the radio beacon emitters carried by the F-105 crew, and overflew the area looking for a safe ingress and exit route for the slower, more vulnerable helicopter.

As the SANDYs examined the area where the two crewmembers were located, several of them received hits from .51 caliber rounds. The weather was becoming bad and was forecast to become even worse over the next few days. Major Clark knew that the pickup had to be made quickly or the weather would close in minimizing the possibility of a successful rescue. To find an ingress path for the choppers, Major Clark flew up one valley and down another until he found a way in. He found one valley with a ceiling high enough for the choppers. The valley was narrow. In order to stay within the valley the A-7s had to hold a continuous two-G turn. One of the HH-53s was able to home in on the SANDY's orbit and, along with the SANDYs, approached the pickup point from 40 miles to the west. The flight was picking up small arms fire from Captain Thaete's position. In response to the ground fire, SANDY Lead and 02 fired on the guns while SANDY 03 and 04 fired smoke rockets to create a smoke screen for the helicopters.

The Jolly dragged its rear wheels through the grass as it taxied to Captain Thaete's position. The door gunner on the Jolly grabbed Captain Thaete and yanked him into the helicopter. Then he immediately swung the 7.62 minigun into position and began firing on a .51 caliber position firing on the helicopter. This gun duel continued for a few seconds until the enemy gun site was silenced by hundreds of minigun rounds. The HH-53 then quickly picked up Major Maier at the bottom of the hill. The Jolly then headed west toward home at maximum power. The SAR force then withdrew having sustained minimal damage. While covering the HH-53's exit, on Major Clark's last pass against several active enemy gun sites Major Clark's A-7 took a direct hit. The round damaged the A-7, he lost his instruments and electrical systems as he pulled up into the clouds. As he broke out on top he was

joined by the other A-7s. The flight of A-7s headed out to sea in case Major Clark would have to eject from his damaged aircraft. He flew south and landed at Da Nang Air Base in South Vietnam. The two downed crewmembers landed at Nakhon Phanom RTAFB, and after receiving minor medical treatment returned that after noon to Korat. The 388th Wing Commander congratulated those who participated, "I know how much courage was involved. This was an example of everyone working together in what was probably one of the toughest SARs we've ever had here." For his part in the rescue, Major Clark was awarded the Air Force Cross.

F-4 Losses

During 1972 the 388th TFW lost seven F-4Es and one F-4D. Seven of these losses were combat losses, and one was a loss on landing due to bad weather. The statistics point out the dependability of the Martin Baker ejection seats in the F-4. Out of the eight losses, with 16 crewmembers involved, there were no fatalities; seven crewmembers were rescued and the remaining nine captured, held as POWs and released in 1973.

Of the six F-4Es lost in combat, three crewmembers were rescued and nine were captured and held as POWs, none were listed as Missing In Action. One pilot, Captain John Murphy was shot down twice, rescued the first time and captured the second time.

February 10 - Tiger FAC Shot Down At Dusk

Captain John Murphy and 1LT. Tom Dobson, while flying aircraft 66-371 on a Tiger FAC (fast FAC callsign SEAFOX) mission in southern Laos, were hit by ground fire and forced to eject. Both crew members were rescued and returned to Korat.

June 8 - John Murphy At The DMZ, Shot Down Again

Captain John Murphy and 1LT Larry Johnson were flying aircraft 67-303 on a Tiger FAC mission along the DMZ which separates North Vietnam from South Vietnam. Their aircraft was hit by ground fire and they were forced to eject. Both crew members successfully ejected — Lt Johnson landed in the water just off the beach, and Captain Murphy landed on the beach. Captain Murphy was captured by enemy forces on the beach. Johnson paddled further out to sea and was rescued by a Navy helicopter.

July 1 - Paul Robinson over Kep Airfield

While flying a MiGCAP in the vicinity of Kep airfield, BULL flight of four F-4Es was heading east towards the ocean. I was flying in the number two aircraft and Major Paul Robinson

62-4424, seen here with its tail removed for an engine change, was shot down by a MiG-21 over North Vietnam on May 11, 1972. Both crewmembers were captured and held as POWs.

67-277, seen here in February 1972, was downed on July 1, 1972 by two SA-2 SAMs scoring direct hits on the aircraft. The crew Major Robinson and Captain Cheney ejected and were captured unharmed by the North Vietnamese spending the remainder of the war as POWs.

69-276 seen here on the boom during an early morning pre-strike refueling, was credited with a MiG-21 on October 8, 1972 and shot down by a MiG-21 on October 12, four days later.

and Captain Kevin Cheney were in the number three aircraft (67-277). While passing over Kep Air Field the flight received numerous SAM warnings on RHAW (Radar Homing and Warning) equipment. These warnings were caused by the defenses around Kep. Major Robinson's aircraft was flying on my right about 2,000 feet away. Even though the flight was manoeuvering and employing ECM, an SA-2 missile detonated next to Major Robinson's aircraft, rupturing the fuselage fuel tanks. As the fire spread forward towards the cockpit, I saw the canopies jettison and both ejection seats, in sequence, leave the aircraft. Immediately after the pilot's seat left the aircraft, a second SA-2 impacted the aircraft in the vicinity of the rear cockpit. Both Major Robinson and Captain Cheney communicated with the rest of the flight by radio while they were in the floating down in their parachutes. I heard Major Robinson jokingly tell Captain Cheney to meet him at Base Operations at Kep Air Field. They were both captured and spent the next nine months as POWs.

July 5 - Double F-4E Shoot Down

My last mission started early in the morning of July 5, 1972, about 2:00 AM when my pilot, Captain "Nordie" Norwood (also the squadron flight scheduler) awoke me from sleep with a phone call. Captain Norwood and I had been scheduled for two days off (July 4th and 5th) and I had just gone to bed following Independence Day celebration at the O-Club. Nordie told me that Major Bill Elander's backseater had gone DNIF (duty not involving flying) due to a cold, and asked me if I wanted to fly backseat for Major Bill Elander on the "north go" in the morning. Bill, my flight commander, was an excellent pilot and knew the F-4 well, having just come from the Thunderbirds where he had served as Material Officer and pilot of the Number-6 aircraft. I told Nordie I would fly and asked him for the details.

The TOT (time over target) was scheduled for 9:30 AM with takeoff at 7:30 AM. Flight planning and briefings started at 4:00 AM. Major Elander and I would be flying as BASS 04 the fourth aircraft in a flight of four F-4Es, all crewed by members of the 469th TFS. BASS flight was one of four F-4E four ship formations which made up the strike force from the 388th TFW. The first three four ships were the strike aircraft and were armed with 12 MK-82 500 pound bombs each. BASS flight was the Strike Escort flight, with each aircraft loaded with two AIM-7s in the aft Sparrow wells; four AIM-9s, two on each of the inboard pylons; and 20mm ammo in the nose. The two forward Sparrow wells were occupied with ECM jamming pods. All 16 of the aircraft had chaff bundles loaded in the speed brakes. Opening the speed brakes in flight caused the chaff to fall out and deploy.

The 16 aircraft took off and flew east over Laos, South Vietnam, and performed a rendezvous with six KC-135 tankers over the Gulf of Tonkin. After receiving fuel, the aircraft formed up into the ingress formation of four rows of four aircraft each, the spacing between each row was about 1000 feet and the spacing between rows was about one half mile. The 12 strike aircraft were in the first three rows, with BASS flight in the "cannon fodder" position across the rear. Our aircraft (67-339 assigned to the 34th TFS with JJ tail code), flying as 04 was on the outside right corner of the formation.

A low cloud bank obscured the ground below as we crossed the eastern coast of North Vietnam, due east of Hanoi. With the top of the clouds at about 5,000 to 6,000 feet, we remained at an altitude of around 20,000 feet to give us plenty of clear air below to allow visual acquisition of any SAMs which might be fired at us. We were receiving no activity on our RHAW gear and as a result the ECM pods were still in standby. Things all seemed uneventful as we continued our ingress to the target.

"BASS 02's been hit" broke the radio silence. We looked to our left and saw BASS 02 (67-296 assigned to the 469th TFS with JV tail code) going down as Lieutenant Brian Seek, the WSO ejected followed immediately by Captain Bill Spencer the pilot. I remember thinking that at least they were able to get out of the aircraft.

About 30 seconds later, "BASS 04 break right" came over the radio, an immediately I felt a hard bank and pull to the right, followed by a strong bump, similar to bumps felt in rough air. I scanned the instrument in the backseat and noticed the engine tachometer readings. No. 1 was at 0% and No. 2 was windmilling at around 28%. I looked over at the circuit breaker panels above the right side console, and noticed the white collars, which indicate when a circuit breaker had popped (opened), were visible on most of the circuit breakers. Bill's voice came over the interphone, "We've been hit!, I'm trying an airstart."

I replied, "Roger, I'm holding in the ignition circuit breakers, turn right 180 for the shortest distance feet wet", as I reached over to and held in the circuit breakers for the No. 2 engine ignition. Since the aircraft had been trimmed for an airspeed of about 420 knots, and with neither engine supplying thrust nor hydraulic pressure for the flight controls, the nose lowered to the dive angle which would hold 420 knots. Bill later said that he had applied right stick to start the turn, watched the System 2 hydraulic pressure drop to 0, then watched the spoilers on the right wing slowly deploy as the aircraft banked into the turn.

As we were turning I looked over to my left and saw a North Vietnamese MiG-21 flying about 200 feet off our left wing. The aircraft was painted with a transparent medium green paint which allowed the silver tint of the aluminium to show through. As soon as the MiG pilot saw me look at him, he broke left and away. Covering the bottom of each wing was a large red rectangle with a five pointed star cutout in the center. I later determined that these markings represented the North Vietnamese flag. I looked back at the engine instruments and noticed that the No. 2 engine rpm had not changed. Bill called over the interphone "I think we'll have to eject, get ready."

As I looked at the altimeter I replied, "Roger ready to eject." We were just starting to enter the clouds and the altimeter reading was passing 7,000 feet. I heard the canopy jettison and felt the cockpit depressurize as it left. Passing through 5,000 feet I saw the cockpit around me blur as the seat fired from the aircraft. It had been just a little over 20 seconds since we had taken the hit.

I felt the seat fall away and looked up and saw the parachute blossom above me. The parachute canopy was properly inflated so I made the modification to the parachute canopy which releases four lines at the aft of the canopy, making it more controllable by giving the parachute forward velocity. I looked around, saw Bill in his parachute about one half mile away. I also saw the fire ball caused when 339 hit the ground. I removed my survival radio from my vest and made a radio call on Guard channel, "BASS 04B in the chute, I have a visual on 04A with a good chute."

I looked down and saw I was getting close to the ground, so I put the radio away and started to steer the parachute towards a small stand of trees. The rest of the area was rice paddies, and I could see the farmers looking up and pointing at me. I figured my best chance was to land in the trees. As I hit the trees the parachute caught first swinging me like a pendulum below. As I swung forward, the parachute broke loose from the tree allowing me to fall about twenty feet to the ground, landing flat on my back. I got out of my parachute harness, and looking around realized I no longer had my helmet. I had lost it on ejection. I felt a sharp pain in my right shoulder, realizing I had dislocated it, I pulled my arm against my chest which popped my shoulder back into its socket. It was 10:30 AM.

I moved away from my parachute looking for a place to hide. I could hear the barking of dogs and the farmers coming to look for me. About 14 men with rifles surrounded me. The first thing they took was my Seiko watch, followed by my 38 pistol and the rest of my survival gear. I was marched to a village near by and put in one of the huts.

After about two hours I was removed from the hut by about six men in uniform (members of the Peoples Militia) and told through sign language and by the soldiers pointing with their guns to start walking down the trail. While I was walking a couple of teenage girls with guns joined in, a camera was brought out, and the typical propaganda pictures of the girls marching the captured American were taken. As the march continued, my shoulder dislocated again. Again I reset it. The soldiers could see that my shoulder was quite painful, so one of them produced a First Aid kit, took a hypodermic syringe out of the kit and proceeded to give me a shot of pain killer. At dusk we arrived in a small village. I was again placed in a small hut and given a cup of hot tea and a small bowl of rice. I was kept in the hut all night.

The next morning the six uniformed men had been replaced by two officers and about 15 soldiers from the North Vietnamese army. We marched for about an hour until we arrived at another village. This village was on a road and two army trucks were waiting in the center of the village. I was blindfolded and put in the back of one of the trucks. We drove most of the day. In the late afternoon we arrived in Hanoi at the prison called the Hanoi Hilton by the U.S. POWs held there.

I spent nine months as a POW being kept in both the Hanoi Hilton and the "show camp" called the Zoo. I was released along with Bill Elander, Bill Spencer, Brian Seek and other POWs in the last group of U.S. POWs to leave Hanoi (released on 29 March 1973).

In discussions with the other members of BASS flight, and based on comments made to me by the North Vietnamese, the tactics used by the MiGs to shoot us down were very simple. The two MiG-21s, directed by a ground radar site, flew a heading, 180 degrees different from the strike force, which aimed them directly at the strike force. They were at a very low altitude, below the clouds, and therefore could not be detected visually by the strike aircraft. The red flags I had observed painted on the bottom of the MiG's wings were there to identify the aircraft as North Vietnamese to prevent ground gunners from accidently firing on their own aircraft. The MiGs passed under the strike force at high speed, and then performed an Immelmann and rolled out behind the strike force.

67-296, seen here loaded with MK82's and CBU's, was flying as Bass 02 on July 5, 1972 when it was shot down by a MiG-21 as part of a double shoot down. I was flying in 67-339, Bass 04, and was shot down also.

KORAT AIRCRAFT LOSSES - 1972

AIRCRAFT TYPE	SERIAL NUMBER	DATE	UNIT (CALLSIGN)	LOCATION	CAUSE	CREW
A-7D	71-310	December 24	353 TFS (SLAM)	Northern LAOS	Combat Mid Air	Cpt. Noice Rec
EB-66C	54-466	April 5	42 TEWS (BAT 21)	SOUTH VIETNAM Near the DMZ	Combat - SAM	Maj W. Bolte - MIA LTC I. Hambleton - Rec
EB-66E	Unknown	December 24	42 TEWS	KORAT	Landing Accident	Unknown 3 Killed
F-105G	63-8333	February 17	1/WWS (JUNIOR)	Hit In NORTH VIETNAM, Crashed offshore	Combat - SAM (Probable)	Cpt J. Cutter - POW Cpt K. Fraser - POW
F-105G	63-8342	April 15	561WWS (SUNTAN)	NORTH VIETNAM	Combat - SAM	Cpt A. Mateja - MIA Cpt O. Jones - MIA
F-105G	62-4424	May 11	17WWS (ICEBAC)	NORTH VIETNAM	Combat - MiG-21	Maj W. Talley - POW Maj J. Padgett - POW
F-105G	63-8347	May 17	561 TFS	KORAT	Landing Accident	Unknown 2 Rec
F-105G	62-4443	July 29	17WWS (EAGLE)	Hit in NORTH VIETNAM, Crashed offshore	Combat - Own Missile	Maj T. Coady - Rec Maj H. Murphy - Rec
F-105G	63-8360	September 17	17WWS (CONDOR)	Hit in NORTH VIETNAM, Crashed offshore	Combat - SAM	Cpt T. Zorn - KIA 1lt M. Turose - KIA
F-105G	63-8302	September 29	17WWS (CROW)	NORTH VIETNAM	Combat - SAM	LTC J. O'Neil - POW Cpt M. Bosiljevac - MIA
F 105G	63-8359	November 16	561 TFS (BOBBIN)	NORTH VIETNAM	Combat - SAM	Maj N. Maier - Rec Cpt K. Thaete - Rec
F-4D	66-265	July 20	35TFS, 3TFW (SCUBA)	Hit in NORTH VIETNAM, Crashed offshore	Combat - AAA	Cpt J. Burns - Rec 1Lt M. Nelson - Rec
F-4E	66-371	February 10	34TFS (SEAFOX)	Southern LAOS Tiger FAC	Combat - Ground Fire	Cpt J. Murphy - Rec 1Lt T. Dobson - Rec
F-4E	67-303	June 8	34TFS (SEAFOX)	Hit in NORTH VIETNAM, Crashed offshore	Combat - Ground Fire	Cpt J. Murphy - POW Cpt L. Johnson - Rec
F-4E	67-277	July 1	34TFS (BULL)	NORTH VIETNAM	Combat - SAM	Maj P. Robinson - POW Cpt K. Chenney - POW
F-4E	67-296	July 5	469TFS (BASS 02)	NORTH VIETNAM	Combat - MiG-21	Cpt W Spencer - POW 1Lt B. Seek - POW
F-4E	67-339	July 5	469TFS (BASS 04)	NORTH VIETNAM	Combat - MiG-21	Maj W. Elander - POW 1Lt D. Logan - POW
F-4E	67-385	September 22	34TFS	THAILAND	Landing Accident	Unknown 2 Rec
F-4E	69-276	October 12	34TFS (SPARROW)	NORTH VIETNAM	Combat MiG-21	Cpt M. Young POW 1Lt C. Brunson - POW

POW = Prisoner Of War,   Rec = Recovered/Rescued,   MIA = Missing In Action,   KIA = Killed In Action

67-385, seen here with a load of MK82s, was destroyed on landing at Korat during a thunderstorm on September 22, 1972.

The MiGs then closed in to Atoll missile range, staying in the F-4's blind spot (an F-4 aircrew cannot see another aircraft as long as it remains below and behind them). Using visual references to determine when they were in range they fired their missiles using the engine exhaust heat from the back row of F-4s as the target. After the missiles were fired, the MiGs "unloaded" and accelerated down and away from the flight. By the time the MiGs were detected by the strike flight they were supersonic and about 10,000 feet below the strike force. In this manner the two MiGs were able shoot down BASS 02 and BASS 04 and escape safely.

July 20 F-4D - Hit By AAA, Lost At Sea

SCUBA flight was bombing the fuel storage at Kep airfield with MK-82s (500 pound bombs) when it was hit by AAA. On fire, F-4D 00-205 immediately headed towards the coast to get "feet wet." The crew, Captain J. Burns the pilot and 1LT M. Nelson the WSO, ejected over the water and were picked up by Navy helicopters.

September 22 - Lost On The Runway

F-4E 67-385, while landing in a thunderstorm, became uncontrollable, departed the runway and was destroyed. The WSO ejected just prior to the aircraft departing the runway. The pilot stayed in the aircraft. Neither received any major injuries.

October 12 - A MiG Killer is Lost

Captain Joe Young and with 1LT Cecil Brunson in his backseat, flying as part of SPARROW flight on a strike escort mission flying southeast of Hanoi, were attacked by a MiG-21. 69-276 was hit by an Atoll missile. Both crewmembers ejected, were captured and spent the rest of the war as POWs. 69-276 which had shot down a MiG-21 with its 20mm nose gun just four days earlier was lost.

PATCHES